Livelihood Sustainability through Agro-biodiversity Conservation

The Authors

Dr. C. Cinthia Fernandaz, Assistant Professor of Extension in the Department of Forest Resource Management, Forest College and Research Institute, Tamil Nadu Agricultural University, Mettupalayam. She has eight years of experience in the Centre for Agricultural and Rural Development Studies (CARDS), TNAU, Coimbatore and the Krishi Vigyan Kendra, Sirugamani, Trichy in the specialized areas of Agricultural Extension, Forestry Extension, Technology Transfer, Information and Communication Tools, Gender issues in Agriculture and Rural Development and Environmental Extension. Area of research in post graduate and doctoral programme was Gender issues. She was awarded with the President of India Award for the Best Doctoral Research for Poverty Alleviation by the Tamil Nadu Agricultural University, Coimbatore in 2010. She had a brief of experience as Field Officer, Spices Board, Cochin. She has four years of experience in teaching, research and extension. She has co-authored two books, more than 15 research articles in journals of National and international repute. She is currently working in the specialized area of Forestry Extension and Environmental Education through Teaching, Research and Extension.

Dr. R. Sendilkumar, Associate Professor (Agrl. Extension) in the College of Cooperation Banking & Management,KAU, Vellanikkara-P.O. 680656, Thrissur, Kerala. He has 14 Years of service in Teaching, Research & Extension with specialization in Agricultural Extension, Agribusiness Management. He is holding active membership in Indian Society of Extension Education, IARI, Society of Extension Education, TNAU, Coimbatore, Society of Extension Education, Agra, International Society of Extension Education, Nagpur. Journal of Agricultural Extension Management, MANAGE, Hyderabad. His research includes Risk management Education, Agricultural Extension Management, Agribusiness Management. He has Guided 12 Post graduate students of MBA (ABM) and 10 Undergraduate students Project of B.Sc. (Hons) C&B. Has published several research papers in professional journals and presented papers in national and international seminars.

Dr. R. Jude Sudhagar, Assistant Professor of Forestry in the Department of Silviculture and Agroforestry, Forest College and Research Institute, Tamil Nadu Agricultural University, Mettupalayam. His area of research in post graduate and doctoral programme was Tree improvement. He was awarded Dr.Krishnamurthy medal for the Best Post Graduate student in Forestry in 1999. He had a brief stint as Assistant Professor in College of Forestry, Allahabad Agriculture Institute (Deemed University), Allahabad, Uttar Pradesh. He has 10 years of experience in teaching, research and extension. He has contributed to R & D in the field of biofuels. He has authored more than 20 research articles in journals of National and international repute.

Dr. R. Ganesan, Former Dean, Gandhigram Rural University, Gandhigram, TN, India(1996-98) and Former Dean, Agrl. College & Research Institute, TamilNadu Agricultural University, Killikulam(Tirunelveli),TN, India(2010-12) has versatile experience spanning over 33 years of his professional service and contributed to the specialized areas of Agrl. Extension, Rural Development, Technology Transfer, Human Resource Development, Agri-Business Development and alike.He visited various countries such as UK, Australia and USA and established professional linkages. He had undergone the FAO sponsored training "Advances in Agrl. Extension" in the university of Reading, UK during 1998.As Director, Agri-Business Development, TNAU, Coimbatore, he pioneered the initiative of establishing the "Business Planning and Development Unit" with the funding support of NAIP, collaborating with ICRISAT, Hyderabad during 2009-10 and instrumental in getting the National Award for the Best Agri-Business Incubator.He has published 2 books and more than 150 articles in the leading research and popular journals.

Dr. P. Durairasu, I.F.S. is a Chief Conservator of Forests and currently working as a Dean, Forest College and Research Institute, Mettupalayam. As a Professional Forester he has a wider experience in managing, conserving and protecting forests and forests genetic resources. He has served as District Forest Officer, Working Plan Officer, Conservator of Forests and Chief Conservator of Forests and has implemented several projects funded by both state and central government agencies. As a Dean, he has introduced specialization programme in post graduate forestry programme and launched several contract farming projects leading to establishment of industrial wood farming in association with wood based industries. He has published several books and articles related to agroforestry, forest management and conservation.

Livelihood Sustainability through Agro-biodiversity Conservation

Authors

C. Cinthia Fernandaz

R. Sendilkumar

R. Jude Sudhagar

R. Ganesan

P. Durairasu

2014

Daya Publishing House®

A Division of

Astral International Pvt. Ltd.

New Delhi – 110 002

ISBN 9789351302698

Published by : **Daya Publishing House®**
A Division of
Astral International Pvt. Ltd.
– ISO 9001:2008 Certified Company –
4760-61/23, Ansari Road, Darya Ganj
New Delhi-110 002
Ph. 011-43549197, 23278134
E-mail: info@astralint.com
Website: www.astralint.com

Laser Typesetting : **Classic Computer Services**, Delhi - 110 035

Printed at : **Replika Press Pvt. Ltd.**

PRINTED IN INDIA

Acknowledgements

I solemnly bow down before all pervasive, reverent and adorable 'Almighty' who is the total conscious force governing me.

I satiate my profound thanks to Dr. P.Durairasu, Dean (Forestry), Forest College and Research Institute, Mettupalayam, Coimbatore for his warm and willing counseling and valuable suggestions in the development and refinement of this book.

I wish to record my deep sense of gratitude and boundless pleasure to Dr.R.Ganesan, Professor, Department of Agricultural Extension and Rural Sociology, Tamil Nadu Agricultural University, Coimbatore, for his expert guidance, learned counsel and continuous encouragement bestowed throughout the period of building the book, without which this manuscript would not have been possible. I also express my indebtness to Dr. A, Sakunthalai, Programme Coordinator, KVK, Sirugamani, Dr. M. Anjugam, Professor and Head, Department of Forest Resource Management, Forest College and Research Institute, Mettupalayam and Dr.Ravi Kumar Theodre, Professor and Head, Department of Agricultural Extension, TNAU, Coimbatore for their able guidance, constructive criticism and incessant inspiration.

I bestow my sincere thanks to my friends Mr. R. Sendil Kumar, Mr. S. Parthasarathi, Mr. M. Israel Thomas and Mr. C. Karpagam for being the great source of inspiration and individual care rendered for building the book.

I place my ineffable and glowing gratitude to my beloved parents Mr. A. F. Cruzmuthu and Mrs. A. Arokiaselvi for their mellifluous love and affection nourished on me. I thank my husband Mr. A. Britto, daughters Rima & Riya, for their love and care which inspires and helps me abide through every phase of my life.I sincerely thank all the respondents who co-operated and contributed for this study.

I profoundly thank the co-authors for their effort taken to complete the book in time. I also acknowledge the scientists who have been provided their technical expertise through chapter contribution.

I am always indebted to The Tamil Nadu Agricultural University for permitting me to undertake the study and providing the platform to excel. I appreciate and acknowledge the efforts of Astral International Publishers'in execution of the book.

Dr. C. Cinthia Fernandaz

Preface

The world todays rounds behind the Biodiversity and its conservation. It has been realized that there is no life beyond biodiversity. This attempt on Livelihood Sustainability through Agrobiodiversity Conservation is to document and index the level of participation of the Tribal women in the conservation. Their day to day activities in one way or the other involve them directly or indirectly on the Agro Biodiversity Conservation measures.Tribal are the dwellers of forest and have a positive relationship with the nature of the forests. Their traditional and scientific knowledge on the natural resources, environment, agriculture and the biodiversity have been developed from generations together. They also produce a high level of interrelationship between the natural resources and the sustainable livelihood.

The world's biological diversity is a vast and undervalued resource. The comment applies to the agricultural diversity. The word 'Biodiversity' is frequently appearing many times in the headlines of leading newspapers, journals and broadcasting news items with a serious emotional warning. Biodiversity would simply mean "varieties of species right to existence". It also would mean that no one has the right to threaten a species right to existence. Man, being the homosapien should restrain from actions that would erode species but do everything to protect them from extinction. This will not only protect the environment but also enable smooth human survival as well. Thus, all species are equally important for the sustenance of all human and sub-human species. Hence, there is an urgent need to strengthen the food chain of all living creatures with sustainable and judicious use. Thus, the mantra of the century should naturally be "protect species or perish."

To find out suitable and concrete evidences or facts to the above mentioned views, a study was needed. Systematic study of any kind would bring valuable information. The results of the empirical approaches could be used in making necessary changes in the agro biodiversity conservation among the tribal women. With this background, the present study was designed with the title "Participation Pattern of Tribal Women in Agro Biodiversity Conservation and Postulating Environmental Education Module for Sustainable Livelihood", with the following specific objectives to study the perception of the tribal women on agro biodiversity conservation, to examine the participation of the tribal women in agro biodiversity conservation practices, to analyze the livelihood sustainability of the tribal women through agro biodiversity conservation, to design an environmental education module for sustainable livelihood through agro biodiversity conservation.

The Nilgiris is one of the "Biodiversity Hotspots" as declared by the National Biodiversity Authority. The green carpet of crops and natural settings always speaks about the various biodiversity of the district. Hence it was decided that The Nilgiris would be the apt place for the study. The tribal population of the district was also high. As Irula tribe has the largest population followed by the Kothars denoted as agricultural tribes with their main occupation as agriculture, they were purposively selected for the study. The Nilgiris district comprised four blocks. *viz.*, Udhagamandalam, Coonoor, Kothagiri and Gudalur. Majority of Irulas and Kothars lived in Kothagiri block; hence it was purposively selected for the study.

A sample size of 200 tribal farm women was fixed for the study. Proportionate sampling technique was followed for the selection of respondents. Ex-post facto design was selected as an appropriate research design.

Participation of tribal women in agro biodiversity conservation and sustainable livelihood through agro biodiversity conservation were considered as the dependent variables in the present study. Age, educational status, experience in farming, social participation, information seeking behaviour, innovativeness, economic motivation, progressiveness, cosmopoliteness, attitude towards agro biodiversity conservation and decision making behaviour, were considered as independent variables to study their relationship and influence on the dependent variables.

The component wise participation index was measured and analysed in order to gain a full understanding on the participation of the respondents in agro biodiversity conservation practices. Among the four components under which the participation was studied the respondents highly participated on the conservation of genetic resources. The participation of Kothars were somewhat higher that the Irulars. Next to genetic resoures agro biodiversity systems were high among the respondents. While considering the ***ex situ*** conservation components Kothars were superior in conservation activities while compared to the Irulars. The four variables such as information seeking behaviour, decision making, age and cosmopoliteness had significant contribution on the participation of the respondents in agro biodiversity conservation.

Among the twelve components studied seven components influence the livelihood sustainability. These components were food sustainability, nutritional security, soil and water sustainability, economic sustainability and employment generation. The constraints were technological constraints, transfer of technology constraints, physical constraints, economic constraints, personal constraints and marketing constraints.

There were nearly twenty three suggestions among which providing timely environmental education was the foremost suggestion given by all the respondents. Hence awareness campaigns and exhibitions could be arranged to impart the importance of agro biodiversity conservation. Training on recent technologies of agro biodiversity conservation practices and knowledge on technological management were the other important suggestions.

The empirical evidences from the study were taken to develop a module on environmental education for sustainable livelihood. It would be utilized for the various stakeholders such as school students, college or university levels and NGOs. The main theme of this module was agro biodiversity conservation and livelihood sustainability.

While an attempt has been made to provide self – contained standard material in the book, suggestions for improvement will always be welcome.

Dr. C. Cinthia Fernandaz

Dr. R. Sendilkumar

Dr. R. Jude Sudhagar

Dr. R. Ganesan

Dr. P. Durairasu

CONTENTS

Chapter 1
Introduction

"This universe is the creation of the supreme power meant for the benefits of all this creations. Individual species must, therefore learn to enjoy its benefits by forming part of the system in close relation with other species. Let not only one species encroach upon the other's rights".

—***Ishopanishad***

"Nature is an essential partnership", essential because each species has its space, role and performs a function essential to the whole; a partnership because the living components of nature, the species, can thrive and survive only together, because together they create a dynamic equilibrium. Nature is a dynamic activity that is never the same; that changes, and then adapts which involves in equilibrium that remains, unchanged because it always accommodated evolution and biodiversity.

The world's biological diversity is a vast and undervalued resource. It comprises every form of life from the tiniest microbe to the mightiest beast and the ecosystems of which they are a part. It provides humanity with cornucopia of goods and services, from food, energy and materials to the genes which protect our crops and heal our diseases. Only a fraction of one per cent of the world's total species has been properly studied for its potential values to humanity as food, medicine and in industry. People have to be educated about the socio-economic value of biodiversity and that is how our own existence depends upon.

FAO (2006) reported that we know the price of everything but value of nothing in the nature. The comment applies to the biological diversity. The word 'Biodiversity' is frequently appearing many times in the headlines of leading newspapers, journals and broadcasting news with a serious emotional warning. It is widely debated word in current environmental discussions and has widely captured the attention of world community. No doubt, biodiversity is of great importance for human existence, as most of the food production, medicinal drugs, and raw materials for shelter are obtained from biological life that exists around us. Human activities of greed and unwise exploitation of this natural wealth is now in danger. The danger is not restricted only to a particular species, but to the entire planet earth. The present rate of extraction of different organisms is the most serious problem of present time.

Nature has been very kind to man. Man with all his powers and prowess, depends totally on other biological entities for his nutritional needs; he is even unable to synthesis certain biochemical's that are essential for his own well-being. That man has survived such serious biochemical limitations attests to his ability to exploit the biological environment to his own advantage. This exploitation began with plants, animals and was also extended to microbes. Even land, sunshine, wind, forests wildlife etc., existed long before man appeared on the scene and he identified and learnt to use these natural gifts indiscriminately.

Biodiversity: Scope and Importance

Nature has varied forms of creation (plants and vegetation (flora), fauna (animals), insects, birds and microorganism etc.) If all these survive, living would be smooth for all the species on earth. Biodiversity would mean that there is need for diverse biological varieties to co-exist without the threat of extinction. Thus, biodiversity would simply mean "varieties of species and right to existence". It also meant that no one has the right to threaten a species right to existence. Man, being the homosapien should restrain from actions that would erode species but do everything to protect them from extinction. This will not only protect the environment but also enable smooth human survival as well. Thus, all species are equally important for the sustenance of all human and sub-human species. Hence, there is an urgent need to strengthen the food chain of all living creatures with sustainable and judicious use. Thus, the mantra of the century should naturally be "protect species or perish".

In the nature's scheme of things, all species play complementary and supplementary role for the survival of all living entities. One life supports the other, which is an unalterable law of nature, thus all species have dual role; (a) every species survives on other species, and (b) every species adds to the food basket of other species as well. In this sense, all species on this planet could be deduced as 'food basket'. Planet earth is unique in many ways, which is the only sheltering innumerable species to co-exist though, at the same time, each of them is feeding or nourishing one another. Without these diverse varieties putting-up together, a few sacrificing in favour of several others, living would have come to an end. It is for man's smooth existence that he should strive for the maintenance of biodiversity order.

The unlimited greed of man is threatening the healthy and diverse biological order that needs immediate correction. He is exploiting both biological and other precious natural resources beyond the replenishment levels besides polluting them badly, thereby poising greater danger to the future generations smooth survival. In an effort to save erosion of natural resources, several threatened regions are declared as 'bio-diversity hotspots' which includes the Western Ghats and the Eastern Himalayan that are treasure houses for several flora and fauna. It is time to make man realize the negative impact of over drawing or polluting of nature on 'biodiversity' and make him act as a 'trustee' or 'conservator' of nature in true sense. Otherwise, a day may not be far off when our future generation would be compelled to face several risks including shortage of food and other requirements. The time is most appropriate to realize importance of biodiversity conservation for the good of all species. It also calls for a well planned strategy to adequately compensate people involved in biodiversity conservation by foregoing benefits of opportunity costs accruing from commercial crops or by exploitation of natural resources.

Need for Biodiversity in Crops

A few decades ago, the planet earth was very rich with numerous crop, plant and sub-human species. Similarly, there were diverse forms of food habits practiced among different social groups. These varied food habits were essentially sustaining multiple crop varieties too. Over the years, more and more people are leaving their traditional food habits. For example, a large number of ragi (finger millet) eaters are drawn towards rice. Similarly, a large number of bajra and jowar eaters have taken to wheat or rice. Tapioca, which was consumed by certain groups in Kerala in the past, has become a taboo among the elite class. Ragi, horse-gram, jowar, maize, tapioca, bajra etc. are considered as poor men's food basket. Thus, over the years, many people have taken rice or wheat more as a symbol of status. When it so happens, naturally there would be lesser demand for various traditional coarse varieties of crops and thus, such traditional crops varieties would naturally be replaced by a few other HYV of rice/wheat varieties (either cash or plantation or popular food crops).

Consequently, we have lost several useful traditional crop varieties. The Government's remunerative support price for wheat, rice, sugarcane, pulses and oil seeds and the absence of good support price for millets and other crops have drawn a large number of farmers to shift away from growing such traditional crops. To compound the problem further, the Government has been supplying only wheat or rice at subsidized rates through Public Distribution System (PDS) which compels more and more people to take wheat or rice consumption. Since the PDS issue rice and wheat they are forced to change their consumption pattern.

Under the prevailing situation, the demand for millets is declining. Thus, the growers as well as consumers of millets (sorghum or bajra) are not receiving adequate encouragement (both on and off the field). The state policies have therefore led to depletion of several useful millet varieties from the fields, which in turn have been causing loss of many crops. If we want to prevent erosion of such crops, it is essential to propagate diverse varieties of food habits of social groups, we should encourage them to continue their traditional food habits and cultural forms too.

Involvement of Tribals in Biodiversity Conservation

India is well known for the diverse varieties of food crops and other plant species. The climate here is ideally suited for varied forms of species to co-exist. The tribals have been the custodians of the rich biodiverse heritage for over 4000 years. The prevalent diverse cultural and social groups with varying food habits from region to region have sustained diverse food crops too systematically. However, in the recent few decades there has been tremendous strain on our biowealth. The risk has been further accentuated with the introduction of hybrid varieties with the practice of mono crop cultivation besides use of large scale chemical inputs. Thus, one can attribute the loss of several traditional or coarse crop species such as vegetables, fruits and food crops alike to the onset of green revolution.

There has been a large scale extinction of very useful species like medicinal plants, food crops, animal species, birds, farm-friendly insects' microorganism, forest species etc. among the tribes. The forests are threatened by large scale felling, encroachments, submergence due to hydro power/irrigation projects, overdrawing or plundering of natural resources by industries, pollution of environment, mining, quarrying etc. Application of chemical inputs such as insecticides, pesticides and manure etc. in cultivation have led to loss of several species in the soil such as beneficial worms, nematodes, predators, parasites, birds, micro-organisms etc.

On farm (*in situ*) crop diversity conservation is easy and cheapest method to maintain, while the gene banks (*ex situ*) are both prohibitive, cost intensive and inefficient. In fact, several useful crop species are lost in gene banks in the developed countries.

Kumary (2008) suggested a cost effective method could be possible only when the tribal community is fully involved not only in growing various varieties of crops but also by consuming them; as both production and consumption are closely interwoven in the sustenance mechanism. The crops in the fields get naturally evolved as they assimilate necessary resistance to the changing environmental variations. There is a need to grow rich traditional varieties in the nature's intricate laboratory as there is no other simple and cheaper alternative to it. So threat to biodiversity can be moderated if not fully prevented only when the communities maintain diverse cropping pattern, revive and grow traditional crops, continue diverse food habits, avoid chemical inputs in cultivation, use bio-friendly farm practices and avoid mono and plantation cropping. The high yielding varieties and plantation crops have already replaced hundreds of coarse and traditional varieties.

Threat to Crop Diversity

As mentioned earlier, the green revolution, in its wake, has brought in mono culture and thereby resulted in the loss of several traditional crop varieties around the globe. The exotic and hybrid crop varieties are only few in numbers. But they have been ruthless in displacing thousands of traditional crop and vegetable varieties. The society has thus been deprived of several useful traditional varieties thereby causing biological and ecological threats. Over 1000 species are getting extinct annually and in India which had over 30,000 rice varieties prior to the introduction of green revolution is now left with hardly about 50 varieties. Thus, the chemical input based hybrid varieties have displaced the biological wealth, considerably (Ramakrishnan, 2002).

In the name of higher yield, farmers were misled to go in for these exotic varieties. Of late, there is growing realization, at some quarters, that the input costs are very high in the cultivation of hybrid varieties. They are seriously threatened by recurring pests and diseases havoc, demand high dosage of chemical fertilizer, pesticides etc. and are susceptible to even minor weather vagaries.

The varieties are dwarf and therefore the farmers lose precious fodder on which the cattle depend for survival. Most of these hybrid varieties are unsuitable to semi-arid conditions. The farmers have to do very intensive cultivation to succeed while is not proved to be economical as it was given to the world by those advocating or propagating the use of miracle seeds. This fact has been proved beyond any doubt.

Cotton, sunflower and silkworm cultivation could propagate the failures of miracle seeds and led several cotton cultivators to penury or peril both in Maharashtra and Karnataka. In the process, man has poisoned the entire biological wealth of nature mercilessly, unmindful of negative impact of these toxic material. The loss on account of erosion of useful traditional crop varieties has caused irretrievable loss to humanity.

The chemical and poisonous residues have polluted the air, water, land and the food crops. While only 10 per cent of these chemicals are used by the plants as much as 90 per cent of them get into water or soil or air. In the case of pesticides less than one per cent actually would reach the target pests while 99 per cent of them contaminate land, water, air, human beings, other animals, birds, bees, moths, wildlife habitat, etc.

By consuming these all the living beings have residues in them which would have detrimental impact on their health. These residues are causing cancerous and various other deadly health hazards among human beings besides threatening survival of many other species. It is essential that the negative effects of these chemicals on various species be revealed to enable people to fully realize their impact on life-support system of nature.

The astounding strides of science and genetic technology have enabled the bio-scientists to tamper with nature (genes). The creations of exotic species have caused great danger to biodiversity. Cloning of species is the latest danger threatening the future shape of things in the hands of biological tempers. Thus, bio-technicians are not only tampering with nature by altering the DNA composition but also appear to assume the role of the Creator. The over adventurism of the homosapiens has caused havoc and ecological imbalance through indiscriminate industrialization and the uncertainty and psychological fear created by atomic/nuclear research. Industry is not only over exploiting natural resources by directly depleting biodiversity but also polluting the air, water and soil which in-turn continue to kill several more essential species.

The negative impact created by industry is enormously hazardous thereby threatening life of the coming generations on earth. Before it is not too late, to organize and mobilize people to articulate in favour of protection of biodiversity. The NGOs, scientists, farmers and others concerned should initiate integrated programmes aimed at preventing further damage to the already ruptured nature. The farmers need to be made aware of the *pros* and *cons* of human adventurism in the name of both technological strides and green revolution. Timely and corrective action programmes to sensitize people on negative impact of commercialization of agriculture are needed to save the future of this planet before we research a point of 'no return'. (Marimuthu, 2001)

Need for Field on Seed Sanctuaries

IISD (2005) reported that the best form of conservation of plant varieties is in the fields, gardens, forests and common lands; thus, these are the true storehouse and laboratory of nature. The farmers, school children, civic society and others concerned should be encouraged to develop mini-plots to identify the endangered species. And grow them on these plots, use their own traditional seeds for sowing and use the excess produce for domestic consumption or sell. Since collection and storage is not a one time process, there is need to create eternal consumption needs for such varieties.

This in turn would impel the farmers to grow such crops year after year, and thereby contribute to the conservation of traditional varieties. Farmers' complete participation is essential in conserving these varieties. Aptly therefore such 'small farms' could be called as 'mini seed sanctuaries'. Such sanctuaries could be enlarged to grow medicinal plants and trees, traditional and threatened crop varieties and propagate use of bio-friendly farm and manure practices. Incentive mechanism should be evolved to suitably reward, appreciate and motivate people to involve in a big way in maintaining rare seed sanctuaries.

Further, a gene kept in a bank is prevented from the cycle of natural evolutionary processes, particularly amidst the changing environmental and ecological conditions. *In situ* preservation, therefore, outweighs all other methods of conservation while *ex situ* conservation could be used as a very last resort. *In situ* biodiversity conservation would enable the seeds to develop compatibility with environmental variations through the evolutionary processes.

Collection/Cultivation Priorities of Threatened Species

The endangered varieties should be enlisted and collected systematically. Such varieties should be grown on several plots. Adequate encouragement should therefore be provided to such mini seed sanctuaries with a view to save the endangered species on farms. Expeditions to collect traditional crop and plant varieties of ecological and medicinal importance should be conducted, from time to time. The importance of a species should not be viewed from the narrow angle of human beings alone, but should be seen in the holistic sense. The schools, colleges, progressive farmers and other interested groups should be integrated in the seed collection expedition. The collected varieties should be distributed for conservation on the farmlands, community/grazing lands, tank/canal beds, kitchen gardens, orchards and around school fields. These would act as an insurance against the endangered species. These should be conserved and used (not over-used) for the collective good of all. Organizing 'seed fares' with good information dissemination would help in collection and exchange of rare traditional species. There is need to evolve an effective mechanism to identify and reward suitably on-farm custodians of rare varieties besides encouraging others to grow more and more of such crops and thereby help preserve them.

Revival of Alternative Practices

There is need to revive traditional organic farming practices, propagate use of bio-pesticides/insecticides, preserve traditional drought and flood resistant plant varieties. We need to popularize diverse varieties of food habits that would in turn help in sustaining diverse on farm varieties too. The adverse effect of chemical inputs in agriculture and the resultant biodiversity loss on account of hybrid crops should be made known to public at large through the electronic and print media. *In situ* seed sanctuaries should be financially supported by development agencies.

Biogas plants and vermicompost units should be adequately financed and spread to rural India in large number which would not only meet the rural energy and bio-compost needs but prevents large scale felling in the community forests. Post agriculture wastes, domestic wastes, vegetable wastes, industrial degradable wastes etc. have vast potential of producing energy, gas and compost. This would help in cleaning the cities and towns too. Use of vermicompost, farmyard compost, and planting of various species to supplement soil fertility, intercropping, etc. need greater propagation. These progressive actions would facilitate saving of endangered species and enthuse people revive the traditional cultivation and preservation practices. The future would be safer in 'bullock-cart' or traditional farming technology than the chemical based cultivation. Let us act as trustees of nature and pass it on, in good condition, to the generations to come. Organic farm practices would be more economical, pollution free, safe and cost effective. If practiced with greater intensity organic and integrated cultivation practices are environmental friendly, which would enable healthy biodiversity order to exist. Ultimately this biodiversity helps people to maintain livelihood. The society gains in many ways by reviving the time-tested traditional crop and farming practices.

The tribes lead a living more close to nature and depend on it for their survival. They should be made aware of the usefulness of the botanical varieties in meeting the diverse needs of nature. Also, they should be educated suitably about the importance of maintaining diversity and the impact of loss of biodiversity. Further, the threatened species should be enlisted and tribes' help is sought in collecting such rare species. Women's involvement in the collection and preservation of these kinds of species is well acknowledged. They should know the value of every species; proper valuing of species is the need of the hour. The development departments should be fully involved in creating necessary awareness and information dissemination about agro-biodiversity conservation among the tribes; especially the tribal women.

To find out suitable and concrete evidences or facts to the above mentioned views, a study was needed. The results of the empirical approaches could be used in making necessary changes in the agro-biodiversity conservation and livelihood sustainability among the tribal women. With this background, the present study was designed and entitled "**Participation of Tribal Women in Agro-biodiversity Conservation and Postulating Environmental Education Module for Sustainable Livelihood**", with the following specific objectives:

1. To study the perception of the tribal women on agro-biodiversity conservation.
2. To assess the participation of the tribal women in agro-biodiversity conservation practices.
3. To analyze the livelihood sustainability of the tribal women through agro-biodiversity conservation.
4. To design an Environmental Education Module (EEM) for sustainable livelihood through agro-biodiversity conservation.

1.1. Scope and Importance of the Study

Eventhough, several theoretical concepts and indices to measure the participation in agro-biodiversity conservation had been identified and proposed worldwide, a very few studies had been conducted empirically to measure the 'livelihood sustainability through agro-biodiversity conservation'. In the present study, a pioneering attempt had been made to develop an index to measure the livelihood sustainability. Accordingly, critical components were identified which were grouped under various dimensions. This index was expected to pave a way for policy makers, planners and developmental professionals.

The components identified in this study would infer as to whether a particular community participate in the agro-biodiversity conservation. The dimensions would facilitate us to know the dimension-wise livelihood sustainability of a particular community. These dimensions were tangible for the policy makers and programme planners. The study on the livelihood sustainability through agro-biodiversity conservation will give a critical idea on the component-wise levels of the community ranging from very low to very high. Based on these details, the policy makers who were interested for the livelihood through agro-biodiversity would take concerted efforts to bring the biodiversity conservation from low to high level.

The identified conservation practices *viz.*, *in situ* and *ex situ* conservation would analyse the basic particulars of the agro-biodiversity. The overall results of the aforesaid practices would provide the following details on all the dimensions to enhance the participation of the clients in the conservation of agro-biodiversity. The genetic resources in the *in situ* conservation provided the details about the traditional seed bank and genetically important crops grown in the field of the clients, which was a vital component in agro-biodiversity. The system in agro-biodiversity gave the information on the important subsystems of the diversity and to what extent the respondents participate in conserving them. Biotic components provided the information on the biotic subsets of the biodiversity in the context of agriculture. Traditional and cultural intervention was studied under the social values which provided the valuable socio-economic base in the participation component.

Livelihood sustainability through agro-biodiversity conservation would help the tribal women to analyze the ways and means for a better living. It provided information in the following subheads such as food sustainability, nutritional security, health management, shelter security, soil and water sustainability, economic sustainability, employment generation, educational security, assets management, cultural sustainability, infrastructure and marketing sustainability and political sustainability. These details supported to develop an index to measure the livelihood sustainability of the elects through agro-biodiversity conservation. This index would be very much helpful to classify a community as low and high in the conservation process. This would also support the government to have a focus to benefit the project designers of the environmental aspects to reach the unreached.

The empirical analysis on profile characteristics with the participation and livelihood sustainability as dependent variable would provide the suitably contributing variables for the study. Furthermore the Sustainable Livelihood Index (SLI) will be an effective tool used to verify the dimensions of agro-biodiversity conservation. The results of case analysis had scope to help the tribal women to know about the ways for participation in the conservation activities. The case analysis will motivate them to build confidence in their day to day responsibilities. Successful tribal women will be treated as a role model and acknowledged by the members of the society.

The constraint analysis would help the planners to eliminate the obstacles and thereby increase the effective participation and livelihood sustainability of the respondents. The empirical model developed using all the Nilgiris experienced in the study may be given due consideration while planning the conservation programmes. The suggested module on environmental education would give an idea on how to execute a programme by seeing various dimensions. In future, it would enhance the participation of the respondents in agro-biodiversity conservation and increase their livelihood sustainability.

1.2. Limitations of the Study

This study was confined to The Nilgiris of Tamil Nadu, which brought out the generalizations for similar areas. This study had no exception to the limitations of time, financial resources and conveyance facilities that would normally be encountered by a student researcher. However, every effort had been put forth to conduct the study as objective as possible.

1.3. Organisation of the Study

In addition to the introductory chapter, four more chapters have been organized in logical order. The second chapter deals with theoretical orientation which was conceptualized based on the review of literature, the third chapter deals with research methodology, which describes the study location, sampling design, identification of components, indices tool for data collection and statistical tests used, while findings and discussion are presented in the fourth chapter. The summary and conclusion of the study with the major findings, implications and suggestions for future area of research are presented in the fifth chapter.

Chapter 2
Theoretical Orientation

Past studies pave way for future research endeavors. Perusal of available relevant literature is essential for any research programme. Therefore, earlier studies were explored and analysed critically in the present investigation in the following sub heads:

2.1. Agro-biodiversity and sustainable livelihood – concepts, approaches and dimensions

2.2. Profile characteristics

2.3. Perception of tribes on agro-biodiversity conservation

2.4. Participation of tribal women in agro-biodiversity conservation

2.5. Livelihood sustainability through agro-biodiversity conservation

2.6. Women and environmental issues

2.7. Constraints and suggestions of tribal women

2.8. Environmental education -approaches and dimensions

2.9. Indices framed on developmental issues

2.1. Agro-biodiversity and Sustainable Livelihood: Concepts, Approaches and Dimensions

Biodiversity and Agro-biodiversity

"Biodiversity" could be explained as variety of lives, where 'bio' means life and 'diversity' means the variety of life forms. There are millions of people who directly depend on biodiversity for their needs of food, firewood, housing, medicine and fodder. Many plants have contributed to life saving drugs. For all of us who live in India, biodiversity has a special cultural meaning.

Eldredge and Salthe (1984) revealed that the concept of biodiversity encompasses the entire biological hierarchy from molecules to ecosystems. It includes the entities recognizable at each level (genes, taxa, communities, etc.) and the interactions between them (nutrient and energy cycling, predation, competition mutation and adaptation, etc.). These entities are heterogeneous, meaning that all members at each level can be distinguished from one another.

Margules (2002) reported that the term "biodiversity" was coined by Walter G Rosen on 21-24 September, 1986 at the National Forum on Biodiversity held in Washington DC. Biodiversity then refers to all biological entities, which is an intimidating idea because it seems impossible to operationalize it. The standard move at this stage is to suggest that three entities capture what is important about biodiversity: genes (alleles), species and ecosystems.

Williams (2005) conveyed that biodiversity surrogates might be species, species assemblages such as vegetation or habitat types, environmental domain or combination of these.

From the above definitions and discussions agro-biodiversity was operationalised as the variety of crops in a particular location with the three entities such as genes, species and ecosystems and also the human dependency on agro-biodiversity.

Levels of Biodiversity

According to Ramakrishnan (2002) theoretically there are three levels of biodiversity but all three are intertwined in such a way that practically it is impossible to separate.

a) **Genetic diversity:** Genetic diversity deals with the variation of genes within a species between individuals of a population or between different population groups of the same species.

b) **Species diversity:** Species diversity deals with the variety of living species in defined geographical area *i.e.*, species richness or species abundance.

c) **Ecosystem diversity:** Ecosystems diversity deals with the difference between ecosystem types, including the habitat diversity and ecological processes in an ecosystem type. Patterns of energy flow and water cycle are distinct in each ecosystem.

The Biological Diversity Act of India, 1998

Brahmi (2004) has explained the objectives of the act are to provide for conservation of biological diversity, sustainable use of its components and equitable sharing of the benefits arising out of the use of biological resources and for matters connected there with or incidental thereto.

Salient provisions of the act were:

Some of the salient provisions made in the Biological Diversity Act (BDA) 1998 for regulation of access to biological diversity, its conservation and sustainable use are:

1. Conservation and sustainable use of biological diversity.
2. Conservation and development of areas important from the stand point of biological diversity by declaring them as biological diversity heritage sites.
3. Protection and rehabilitation of threatened species.
4. To respect and protect knowledge of local communities related to biodiversity.
5. Regulation of access to biological resources of the country with the purpose of securing equitable share in benefits arising out of the use of biological resources and associated knowledge relating to biological resources.
6. To secure sharing of benefits with local people as conservers of biological resources and holders of knowledge and information relating to the use of biological resources.
7. Involvement of institutions of self government in the broad scheme of the implementation of the act through constitution of committees.

Pandey (2004) reported that we do not know how many species are available in the universe. Different projections have been made by a number of ecologists, evolutionists and taxonomists based on different assumptions and yielding desperate numbers. At global level, it is estimated that there exist 5-30 million species of living forms on our earth. These include 3,00,000 species of green plants and fungi, 8,00,000 species of insects, 40,000 species of vertebrates and 3,60,000 species of micro-organisms.

Brahmi (2004) mentioned that the tropical forests were considered as the richest in biodiversity. The reasons may be many. Some of them are:

1. The conditions for evolution are optimum, and for extinction fewer.
2. Because of the low rate of extinction the species diversity was conserved over geological time.
3. The diversity is the result of interaction between climate, organisms, topography, parent soil materials, time and heredity; warm temperatures and high humidity are favourable climatic conditions for many species to survive.
4. The species tend to be stable, continuing their life there itself compared to the dispersal mode of temperate species which are less stable.
5. Tropical areas receive more solar energy over the year. Thus, tropical communities are more productive or greater resource base that can support a wider range of species.

Agro-biodiversity

The Convention on Biological Diversity (CBD) (1998) reported that agricultural biodiversity is a sub-set of general biodiversity including all cultivated varieties. Cultivated varieties can be broadly classified into "modern varieties" and "farmers or traditional varieties". Modern varieties are the outcome of scientific breeding and are characterised by a high yield and a high degree of genetic uniformity. In contrast, farmer's varieties (also known as landraces) are the product of breeding or selection carried out by farmers. They represent higher levels of genetic diversity and are therefore the focus of most conservation efforts. Agricultural biodiversity contributes to food security and livelihood security and underpins the development of all food production. It is the first link in the food chain, developed and safeguarded by farmers, livestock breeders, forest workers, fishermen and indigenous peoples throughout the world.

Ramakrishnan (2002) had reported that although the term "*agricultural biodiversity*" is relatively new, it has come into wide use in recent years as evidenced by bibliographic references—the concept itself is quite old. It is the result of the careful selection and inventive developments of farmers, herders and fishers over millennia. Agricultural biodiversity is a vital sub-set of biodiversity. It is a creation of humankind whose food and livelihood security depend on the sustained management of those diverse biological resources that are important for food and agriculture. Agricultural biodiversity includes:

- ✰ Domesticated crop and 'wild' plants, including woodland and aquatic plants (used for food and other natural resources based products), domestic and wild animals (used for food, fibre, milk, hides, furs, power, organic fertilizer), fish and other aquatic animals, within field, forest, rangeland and aquatic ecosystems.
- ✰ Non-harvested species within production agroecosystems that support food provision, including soil micro-biota, pollinators and so on.
- ✰ Non-harvested species in the wider environment that support food production agroecosystems (agricultural, pastoral, forest and aquatic ecosystems).

Brahmi (2004) reported that agricultural biodiversity, sometimes called agro-biodiversity, "encompasses the variety and variability of animals, plants and micro-organisms which are necessary to sustain key functions of the agroecosystem, its structure and processes for, and in support of, food production and food security". It is the product of sustainable agroecological production systems and these systems simultaneously depend on a wide range of agricultural biodiversity.

Agro-biodiversity Conservation

Anonymous (1996) reported that conservation in its broadest sense means to stop the unplanned development that breaks ecological as well as human laws by careless use of natural resources. In technical terms, we use the term conservation to denote policies and programmes for the long term retention of natural communities under conditions which provide the potential for continuing evolution. Conservation is important to maintain the existence of life on the earth.

Bunning and Hill (1996) reported that the genetic erosion of crop plants by neglect and by the elimination of potentially valuable wild and weedy relatives is a great loss to the future biological research. While pondering on conservation, one cannot forget the fact that whenever a species becomes extinct, many associated beings also meet the same fate. Our planets requirements and services depend mainly on the biological resources, which provide us nourishment, clothing, housing, fuel and medicine but also meet our several other requirements. Therefore, the knowledge of biodiversity is of immense utility in planning sustainable livelihood and conserving the natural resources.

According to Ramakrishnan (2002), the natural causes for depletion of biodiversity include drought, land slides, floods, storms, earthquakes, diseases, predation, fire, etc. Biological extinction has been a natural phenomenon in geological history. The rate of extinction was perhaps one in 1000 years. The artificial causes include mainly hunting, persecution, habitat destruction, overexploitation of biological resources, grazing, industrialization, urbanization, pollution, scientific and educational research, road and dam construction and introduction of exotic plants and animals, expansion of agricultural lands, thinning of ozone layer, global warming and acid rains are also some of the other reasons. Uncontrolled and excessive biotic interference also contributes to its loss. The factors for loss of biodiversity are also being classified under 1) anthropocentric perspective, 2) resource perspective and 3) ecosystem perspective

Ramakrishnan and Chidambaram (2003) discussed the importance of biodiversity conservation as follows: The fundamental questions that need to be addressed for conservation are: Why conserve biodiversity? What to conserve? How rapidly can ecosystem be conserved? What should be the priorities and strategies? How well can a conservation and management project be designed and implemented? These questions are to be answered in the conservation of agro-biodiversity.

Brahmi (2004) had reported that man's interaction has speeded up extinction rates up to one in 10 years and currently it is perhaps one in one year. Rich species diversity is slowly being lost for ever. Man's activities resulted in the extinction of over 23 wild species of plants and animals from India. Indian cheetah, the pink headed duck and mountain quail have already become extinct from the Indian fauna and several species of mammals, 47 species of birds, 15 species of reptiles, 3 species of amphibians, a number of butterflies, moths, and beetles are facing the danger of extinction. 1500 species of vascular plants are considered to be endangered. Illegal trade and introduction and expansion of high yielding crop varieties affect the survival of these endangered plant and animal species. The wild strains of rice, millets, oil seeds, vegetables and legumes have disappeared and three species of cattle, five of sheep, five of camel, four of horses and all varieties of poultry are endangered. Thus the major genetic bases of these plants and animals which could have served as germless banks for desired plants and animals have been destroyed. India's biodiversity is one of the most significant in the world. Of the world's top priority biodiversity hot spots, India has two within its boundaries.

Biodiversity Conservation Strategies

According to Ramakrishnan (2002) in recent years, laudable attempts have been made to identify biodiversity hot spots and unique habitats and to preserve, protect and propagate them through *in situ* and *ex situ* measures.

1. ***In situ* conservation strategy:** This strategy emphasizes on the conservation work at original site of biodiversity *i.e.* in wild situation. Conservation of overall diversity of genes, populations, species, communities and the ecological process comes under this strategy. There are 37000 protected areas in the world in 1994. In India at present there are 448 wild life sanctuaries 85 National Parks and 10 Biosphere Reserves, covering about 4.2 per cent of the total geographical area.
2. ***Ex situ* conservation strategy** : This strategy says that conservation work should be done outside the natural habitat in the form of botanical and zoological gardens, conservation stands, seed and seedling banks, pollen banks, germ plasm banks, tissue culture banks, gene and DNA banks, etc.

Strategy for *in situ* Conservation

Tomar (2007) had given the following: Conservation strategies are proposed based on the nature of threats faced by the species and their habitats. The strategies are of two types: species specific and habitat (area) specific. For effective implementation of conservation strategies, the role of local people and the scope of improvement of their role have been investigated. This would help in effective local peoples' involvement in the conservation of species and habitats. The following were the salient features of *in situ* conservation.

a) Habitat Conservation

In the case of habitat, the strategies are developed based on general threats to the habitats. These include,

- ☆ Strengthen protection/conservation in PAS (Protected areas)
- ☆ Establish Ecological Sensitive Areas – ESA
- ☆ Stop habitat conversion for plantations, agriculture, mining in selected areas
- ☆ Recognize grasslands as natural habitats
- ☆ Plan coastal zone management with due consideration of ecologically sensitive habitats

b) Species Conservation

Conservation options for the species threatened due to their unsustainable harvesting, destruction under traditional beliefs and reproductive limitations of the species include the following:

- ☆ Create awareness among people and forge partnerships
- ☆ Promote conservation and sustainable use (grazing)

- ☆ Encourage cultivation of medicinal plants as feasible
- ☆ Develop non-destructive extraction methods
- ☆ Improve the reproductive capabilities of species under natural or artificial conditions.
- ☆ Cultivation of medicinal and other economics important plants should be promoted. This would increase the availability of these species and also bring higher financial benefits in short time to the people.
- ☆ Develop clear guidelines for the sustainable collection of plant and their parts, based on ecological principles. The guidelines may include :
    - Young and immature plants should not be allowed to be harvested, unless specifically required.
    - Extraction of critically endangered species should be banned till the status improves in the wild.
    - If the parts of the plant are harvestable year round (*e.g.*, bamboo, grass), the harvesting should be discouraged in regeneration stage (mostly monsoon).
    - Leave a minimum of 15 per cent plants intact to effectively reproduce.
    - If the harvesting involves the reproductive parts of the plants (flowers, fruits, seeds) care should be taken that not more than 85% of these parts should be harvested from any plant.
    - Minimise the use of pesticides in cultivation of medicinal / economically important plants as these would be cultivated in the vicinity of the forestland and their excessive use may adversely affect the forest biota.

Strategy for *ex situ* Conservation

The Convention on Biodiversity (1998) reported that the *ex situ* conservation needs to be brought under one authority. It may be under the Sate Biodiversity Board. The authority should work in tandem with authority for conservation and promote the botanical gardens to help conservation of plants; particularly those threatened in wild, and monitor the progress. It may ensure that the botanical gardens work for the objectives for which they have been set up. Botanical gardens in each of the biogeography regions of the state will help conservation of the species in these regions.

The botanical gardens should be strengthened towards developing protocols for reproduction of species with poor germination in natural conditions. The collections of botanical gardens should be exchanged among themselves and with others so as to make the enrichment of collection cost effective. Regular meeting between the official should be encouraged to find out the threatened plants and to facilitate their conservation. Close interaction among the officials of all the botanical gardens would ensure that there is no duplication of work while developing the propagation techniques. National institutions could be roped in.

Sustainable Livelihood – Concepts

Singh (1996) has opined that sustainable livelihoods are a systemic and adaptive approach that links issues of poverty reduction, sustainability and empowerment processes (*e.g.*, participation, gender empowerment, and good governance). The attractiveness of sustainable livelihood lies in its applicability to different contexts, situations of uncertainty and in its capacity as a consultative and participatory process for the cross-fertilization of ideas and strategies between various stakeholders. Those living in extreme poverty and outside the formal labour market, for example, constantly improvise their livelihood strategies due to high uncertainty and limited options. A subsistence farmer in the off-season or during drought becomes a wage labourer and could later revert back to farming when it is time to plough the field.

Singh and Gilman (2004) recorded that surfacing a decade ago in the report of the World Commission on Environment and Development, the idea of sustainable livelihoods began as an approach to maintain or enhance resource productivity, secure ownership of and access to assets, resources and income-earning activities as well as to ensure adequate stocks and flows of food and cash to meet basic needs. It was a reflection of the growing recognition that food security was not merely a problem of agricultural productivity but was a problem of poverty in all its multi faceted dimensions. The 1992 United Nations Conference on Environment and Development (UNCED) moved the concept towards an action agenda especially in the context of Agenda 21, and advocated for the achievement of sustainable livelihoods as a broad goal for poverty eradication focused on the poor who live in marginalized areas. Agenda 21 stated that sustainable livelihoods could serve as an integrating factor that allows policies to address development, sustainable resource management and poverty eradication simultaneously.

Anonymous (2005) had reported that a possible option for a conceptual framework within which to place sustainable livelihood is one developed by the International Institute for Sustainable Development (IISD). The framework integrates the concepts of sustainable development and sustainable livelihoods. It is be conceptualized as a diagram merging two interactive triangles, one representing the cornerstones of sustainable development (economic efficiency, environmental integrity and human well being) and other the showing those of sustainable livelihoods (local knowledge, science and technology, and policy structures). It is opined that elements and issues that make for sustainable livelihoods lie at the critical interface of human environment interactions. Political, cultural, religious, social, economic, biological and geo-physical factors simultaneously interact with and in combination with each other produce a variety of functions, processes and products, which shape the way a community makes a living in a given ecozone. Analysis of these factors allows policy makers and practitioners to formulate appropriate and context specific programmes and projects that aim to promote particular sustainable livelihood systems.

Sustainable Livelihood: Strategies and Approaches

FAO (1999) developed the indicators of change for livelihoods as a whole with the following base:

- **Food Security**: Increase/improvement in yields will result in improved food security through direct consumption of produce and indirectly through sale of surplus. Poor families have food security, *i.e.* have enough stores and income to eat throughout the year increase in amount of daily meals and improvement in quality of meals improved nutritional status will result in improved family health, *i.e.* balanced diet consisting of beans, meat, vegetables, papaya etc. decrease in diseases caused by malnutrition, *i.e.* malnutrition in babies.
- **Income and Expenditure**: Improved vegetable production and field crops results in extra income from sale of surplus. Increase in income (money from sale of surplus) increases ability to pay school fees and to purchase other assets such as livestock, clothing, inputs required for improved or new housing, and vehicle.
- **Employment:** Local employment opportunities increase as a long-term result of farmers' increased income through sale of surplus higher yields. These farmers will employ locally available labour to work higher producing fields, as will have the financial means to do so. Expect this to happen after five years of continued new practices..
- **Status:** Change from poorer status to higher one.
- **Access:** Access to resources, *e.g.* Protected water tanks in all villages.

According to Gari (2001) the consideration of agro-biodiversity and indigenous knowledge results in new modes of agricultural and rural development, which directly address household food security and nutrition aspects, while providing an integral approach to the entire agriculture-food-nutrition-health complex at the rural grassroots. Next there follows a set of strategic components that can empower the rural poor in the face of food insecurity provided adequate changes in policies, in agricultural programmes, and in practices at the farmer level are accomplished.

Singh and Gilman (2004) suggested that the livelihood systems comprise a complex and diverse set of economic, social and physical strategies. These strategies are realized through the activities, assets and entitlements by which individuals make a living. Sustainable livelihoods are derived from people's capacities to exercise choice, and to access opportunities and resources, and use them for their livelihoods in ways that do not foreclose options for others to make their living, either now, or in the future. But sustainable livelihoods are not just about what happens in a given locality. It begins with how people living in poverty, rather than experts, perceive their own reality, and extends to understanding how this reality is related to what happens in the rest of society, to forming new relationships within and outside the locality, and imagining alternative paths of social transformation to the present ones. It stresses the need to support and protect people's capacity to act and produce. Yet people's productive lives are not reduced in narrow economic terms.

2.2. Profile Characteristics of the Respondents

Age

According to Misra and Singh (1998) revealed that 46.4 per cent of the elected representatives of the tribes were between 21-35 years of age which indicated that new leaders are very young, 35.70 per cent were in the age group of 36-45 years, while 17.9 per cent were between 46.55 years of age.

Jamatia (1999) reported that 44.00 per cent of tribal farm women belonged to middle age group, 31.33 per cent young group and 24.67 per cent old age group.

Santha Govind (1999) conducted survey among the tribal women who involved in panchayat raj system in three states. In Haryana, of the total 45 elected representatives, majority (68.89 per cent) of them were 40 years and above, whereas in Kerala, 52.70 per cent and in Tamil Nadu, 80.95 per cent of the elected women were less than 40 years of age.

Rani (2000) indicated that 43.20 per cent of both the elected and the defeated respondents belonged to the age group of 31-40 years, 36.40 per cent were in the age group of 21-30 years, 15.10 per cent were from the age group of 41-50 years and only 5.30 per cent come under the age group of 51-60 years.

Palmurugan (2002) sated that 43.20 per cent of the farm women belonged to middle age group, followed by 36.50 per cent and 20.30 per cent who belonged to young and old age groups respectively.

Sailaja (2002) observed that 95.00 per cent of the farm women were belonged from middle to young age group and only 5.00 per cent in the old age category.

Bhuvaneshwari (2005) revealed that majority of the tribal respondents were (66.00 per cent) of young age group followed by medium and old age group.

Educational Status

Jamatia (1999) reported that 81.33 per cent of tribal farm women were illiterates. Only 13.33 and 5.34 per cent of the respondents were having primary and medium level of education.

Santha (1999) studied the educational level of the elected tribal panchayat members in the sample, it showed the disparity in women's development in three States. In Haryana, only 31.11 per cent were literates and the remaining 68.89 per cent of the elected women representatives in panchayat raj bodies were illiterates. In Tamil Nadu, 88.09 per cent were literates. Of this 25 per cent studied upto primary level, 16.67 per cent upto middle school level, 22.62 per cent upto matric level, 11.90 per cent upto pre-degree level and only 7.14 per cent gained collegiate education. In Kerala, all the sample population was literates. Among the 74 literates, 24 were found professionally qualified in different areas such as teaching or law.

Rani (2000) classified the tribal representatives according to their levels of education. Out of the 206 respondents, 57.80 per cent were found to have studied upto the secondary level, 18.40 per cent upto the primary level, 11.20 per cent upto diploma level, 7.80 per cent had studied upto college level, while less than one (0.90 per cent) were illiterates.

Saravanakumar (2000) reported that 38.03 per cent of tribal women completed their secondary education and equal per cent of them (32.40 per cent) had completed their middle school level education. Nearly 26.77 per cent had completed their primary level of education.

It was reported by the author Palmurugan (2002) that 86.40 per cent of farm women possessed medium level of educational status, followed by 8.10 per cent who had low level of education. The remaining 5.50 per cent were highly educated.

Sailaja (2002) indicated that 43.00 per cent in primary education whereas equal per cent (20.00 per cent) of the farm women were distributed in the illiterates and secondary level of education.

Shanthasheela (2002) stated that 83.57 per cent of the farm women were distributed in the categories of primary to secondary level education.

Bhuvaneshwari (2005) revealed that majority of the tribal respondents who were actively playing role in the conservation of agro-biodiversity were (48.00 per cent) of illiterate in nature. This group was from the Irula community. The next majority group was primary followed by secondary education.

Experience

Madivanane (1990) stated that nearly half (47.00 per cent) of the farm women had low farming experience followed by nearly on equal percentage of the respondents with high (27.00 per cent) and medium (26.00 per cent) level of farming experience.

Manjula (1991) indicated that in the case of tribal farm women nearly half (44.94 per cent) of them had medium level farming experience (11-25 years). She also opined that among the tribal farm women 57.16 per cent were not members in any of the organisations, while 28.57 per cent had low level social participation, 12.24 per cent had medium level social participation and 2.00 per cent of them had high level social participation.

Sujatha (1996) found that 46.66 per cent and 41.67 per cent of farm women had medium and high level farm experience respectively. 11.67 per cent of them had low level experience.

Selvarani (2000) observed that 43.33 per cent of the tribal farmers possessed medium level of farming experience, followed by 36.67 per cent with low level of farming experience. Farming experience was high only with 20.00 per cent of the tribal farmers.

Marimuthu (2001) revealed that 36.00 per cent of the tribal women possessed medium level of farming experience followed by 32.86 and 30.95 per cent with low and high levels of farming experience respectively.

Social Participation

Saraswathi (1993) reported that 41.67 per cent of the tribes had low level social participation, followed by medium and high levels (35.83 and 22.50 per cent respectively).

Premavathi (1997) stated that 53.00 per cent of farm women possessed a medium level of social participation, followed by 33.00 and 14.00 per cent with low and high level of social participation respectively.

Jamatia (1999) revealed that 69.33 per cent of the tribal women had a low level of social participation, followed by 26.00 per cent of the respondents who had medium level of social participation.

Selvarani (2000) found that majority of the tribal respondents (59.17 per cent) were with low level of social participation, followed by 30.83 and 10.00 per cent with medium and high levels of participation respectively. Kothas significantly had more social participation behaviour than Todas.

Marimuthu (2001) revealed that 81.43 per cent of the respondents had medium level social participation and 18.57 per cent had high level social participation.

Bhuvaneshwari (2005) revealed that majority of the tribal respondents (96.00 per cent) possessed low social participation, followed by medium level of social participation.

Information Seeking Behaviour

Senthamarai (1997) stated that less than three-fourth (70.83 per cent) of the farm women had low level of information seeking behaviour followed by medium (30.00 per cent) and high (25.83 per cent) level.

Lakshmana (2000) reported that among the respondents 45.83 per cent had low information seeking behaviour, 45.00 per cent had medium level of information seeking behaviour and 9.17 per cent of respondents had high level of information seeking behaviour.

Selvarani (2000) revealed that more than one-third of the tribal respondents (38.33 per cent) had medium level of information seeking behaviour closely followed by high (31.67 per cent) and low level (30.00 per cent). Toda farmers had significantly high media participation than Kotha farmers.

Marimuthu (2001) found that 60.00 per cent of the respondents belonged to low level information seeking behaviour, followed by 30.00 and 10.00 per cent who belonged to medium and high level information seeking behaviour.

Innovativeness

Parvathi (1995) reported that majority (73.80 per cent) of the farm women were under medium level innovativeness followed by low level innovativeness (18.10 per cent). Only 8.10 per cent of the farm women were under high level innovativeness.

Jamatia (1999) revealed that 40.00 per cent of the respondents had medium level of innovativeness, followed by 43.30 per cent with low and only 10.70 per cent had high level of innovativeness.

Saravanakumar (2000) reported that farm women in general had high level of innovativeness. About seventy per cent of the participants had high level of innovativeness and 19.72 per cent had low level of innovativeness.

Marimuthu (2001) revealed that 50.48 and 49.52 per cent of the respondents belonged to low and medium levels innovativeness categories respectively.

Majority (87.80 per cent) of the farm women ad medium level of innovativeness, followed by 10.80 and 1.40 per cent of farm women who had low and high levels of innovativeness respectively was inferred by the study of Palmurugan (2002).

Bhuvaneshwari (2005) revealed that majority of the tribal respondents of the Irula group (66.00 per cent) had low innovativeness followed by medium and high levels of innovativeness.

Economic Motivation

Senthamarai (1997) stated that 42.50 per cent of the respondents were with low level of economic motivation, followed by medium (32.50 per cent) and high (25.00 per cent) levels.

Kumari (1998) observed that more than 90.00 per cent of the farm women in all the three farming system were found to have medium to high level of economic motivation.

Jamatia (1999) reported that about 36.00 per cent of tribal women had low level of economic motivation, followed by 34.00 and 30.00 per cent of respondents who had medium and high levels of economic motivation.

Selvarani (2000) observed that moderate level of economic motivation was found with more than half (65.84 per cent) of the tribal farmers. Kotas had significantly higher economic motivation than Todas.

Marimuthu (2001) found that 44.76 per cent of the tribals had medium level economic motivation, followed by 33.33 and 21.91 per cent of them belonging to low and high level economic motivation categories respectively.

Bhuvaneshwari (2005) revealed that majority of the tribal respondents (84.00 per cent) had medium economic motivation followed by high and low levels of economic motivation.

Progressiveness

Premavathi (1997) found that 68.00 per cent of farm women had medium level of progressiveness while 27.00 per cent had low and 5.00 per cent had high level of progressiveness.

Senthamarai (1997) stated that about two-fifth (40.83 per cent) of the respondents had medium level of progressiveness, followed by low (31.67 per cent) and high (27.50 per cent) levels.

Jamatia (1999) observed that majority (72.67 per cent) of the tribal farm women had poor progressiveness, followed by 20.00 and 7.33 per cent of the tribal farm women who had high and medium level progressiveness.

Marimuthu (2001) reported that 44.29 per cent of the tribal farmers had low level progressiveness, followed by 34.76 and 20.95 per cent with medium and high level progressiveness respectively.

Cosmopoliteness

Sripal (1983) reported that 75.00 per cent of the tribes were neither fully localites nor fully cosmopolites.

Kumari (1998) indicated that cent per cent of the respondents had low urban contact. Prior to rehabilitation and settlement the Onges were rarely ventured out and were busy and happy with their semi nomadic life.

Marimuthu (2001) revealed that 41.00 per cent of the tribal farmers were localite in nature, followed by 38.10 and 21.42 per cent of them belonged to neither fully localite nor fully cosmopolite and cosmopolite nature categories respectively.

2.3. Perception of Tribes on Biodiversity Conservation

This section deals with various tribal groups of our country and their way of living. The crop cultivation method of these tribes has been discussed to reveal the natural living of them in accordance with the Environment.

A. Kammara (Tribe of Andhra Pradesh)

A study conducted by Pratap (1998) on Kammara tribe of Andhra Pradesh revealed that Kammara (Telugu, *Kammara* = artisan, smiths), the community derives its name from their traditional profession, of working with metals and wood.

The salient findings of his research were summarized as below. The Kammaras are inhabitants of the agency areas in East Godavari, Srikakulam, Visakhapatnam and West Godavari districts of Andhra Pradesh where they are cohabitants with other village communities.

The Kammaras residing in the agency areas, due to the practice of inter-caste marriages, are identified into several sects. These sects referred to as race and known by different titles namely; *Bhag* (tiger), *Bhallu* (bear), *Hanuman* (monkey), *Korva* (sun), *Naga* (snake) and *Surabhi* (cow).

The Kammaras residing in agency areas live by their traditional profession of blacksmith and carpentry. They make rough gardening and agricultural implements and also cheap brass ornaments and jewellery. With the improvements in conveyance and availability of vehicular transport, the traders from outside have brought in improved implements and cheap ornaments to the weekly shandies (weekly market). This has resulted in the Kammaras not being able to sell their old types of wares, forcing them in abandoning their traditional professions to take-up cultivation or to work as agricultural labourers. Those living near the forests adopt the *podu* (slash and burn) type of cultivation while others, eke out a livelihood by collecting small forest produce and selling them in the weekly shandies. During famine (drought) they sustain themselves by living on roots and leaves collected from the forest.

Kammaras celebrate Korrakothu (harvest festival) grandly. *Sambaranivamel* by providing wooden legs and one goat and 10 seers of rice.

B. Paroja (Tribe of Orissa)

Mohanty (1990) investigated extensively on Paroja tribe of Orissa and the findings are given below:

The term "Paroja" is a local Oriya term sometimes pronounced as *Paraja, Parja* or *Poroja*. It appears to be derived from the Sanskrit word Praja which literally means the common people, *i.e.*, the subjects or citizens as distinguished from the former ruling chiefs called the *Raja* or the *Zamindar* in the pre-independent times.

The Parojas use indigenous raw materials like soil, mud, cowdung (*lappy*), bamboo, wood, straw (*piri*) and country made tiles (*jhikar*) for house construction. The walls, veranda and the floors are regularly maintained, by plastering with cowdung and decorating with coloured soils. The Parojas are fond of using red or white coloured soil mixed with cowdung or ashes for colouring the house. Maintenance of the roof is done by thatching the house once a year, preferably after the harvest when paddy straw are available. In order to project the low thatched roofs from the cattle, wattle fencing are made around the house.

The Parojas are seasoned cultivators and agriculture is the mainstay of their economy. Formerly, the Parojas were subsisting upon shifting cultivation, which is now reduced to a secondary or tertiary means of livelihood. Now they mainly depend upon wet cultivation which is supplemented by shifting cultivation and other allied economic pursuits like collection of forest produce, wage earning, animal husbandry, etc.

Their cultivable land is of four major categories. First type is known as *Beda* or wet land, which is best for paddy cultivation. Second category is known as *Huda*, which is plain non-irrigated upland where paddy and minor millets are grown. The third category of land is known as *dongar* meant for shifting cultivation, where oil seeds, millets and pulses are grown. The fourth type is the kitchen garden situated behind the house where different kinds of vegetables are grown. Regular care of the kitchen garden is taken and every year these plots are manured with cowdung.

The major crops grown by the Parojas are of high land variety – namely maize, minor millets, early paddy, ragi, jowar, niger, etc. Recently new crops such as potato, arhar, wheat, sugarcane and vegetables like cabbage, cauliflower, brinjal, tomato, etc., have been introduced by the development agencies. In shifting cultivation sites, they grow minor millets, *kandul, suan* (small millets), ragi, maize, *biri* and early paddy, etc. Fruits and vegetables are also grown in a small kitchen garden behind the house and also in low lands where irrigation is available.

For cultivation, the Parojas use very simple type of agricultural implements namely plough, spade, hoe, axe, etc. Cultivation of land, manuring, sowing, weeding, reaping, harvesting are done carefully and systematically. In all phases of agricultural operations, mutual co-operation and co-ordination among the people are a striking feature.

Rice is their stable source of food. It is supplemented by millets, maize, and vegetables grown by them, while other edible fruits, roots, leaves and tubers collected from the forest around the year. The Parojas are addicted to alcoholic drinks and beverages. Use of liquor is a customary practice in all the magico religious rituals, feasts and festivals. They drink various kinds of liquor such as; mahua liquor, millet liquor (*landha*), rice beer, palmwine (*salap*), etc. *Landha* and *salap* are the most favourite drinks.

C. Bondo (Tribe of Orissa)

Patnaik (1990) conducted a study on tribes of Orissa and reported that the Bondo tribe is divided into three groups the Upper Bondos or Bana-Jangar group who live on the slopes of 3,000 feet high altitude with Mudulipada as its capital in Khairput Block, the Lower Bondos who live at the foot hills under symbiotic pressure of the surrounding peasantry and the Gadaba-Bondos who being greatly influenced by the Gadabas live in the same altitude with the upper Bondos. Both the Bara-angar group and Gadabada-Bondos are considered as hill Bondos.

The Bondos are mainly agriculturists. Besides, they practice shifting cultivation (*kunda chas*) quite extensively. The other sources of their livelihood are gleaning, hunting, fishing, animal husbandry and wage earning.

The Bondos are expert cultivators. They cultivate four types of land-wet land (*jhola orlivang*), up-land (*pada*), hill-slopes (*kunda* or *dongar*) for shifting cultivation and kitchen garden (*danabui*) within the settlement. The clan members collectively own all these types of land other than dongar lands. The irrigated and terraced paddy fields are used for cultivation. They grow paddy by transplanting method. Besides paddy, they grow cereals, pulses, millets and oilseeds such as, *kangu, suan jana*, oat, black gram, maize, *alsi* and niger, maize, tobacco and vegetables are grown in the kitchen garden.

Though forests have been depleted to a great extent, the Bondos as forest-dwellers are still in the habit of going to distant forests to collect bamboo, wood, bark, wild grass, fruits, greens and mushrooms. In leisure hours they trap birds and animals. They also go for ceremonial hunting in-group in the month of March-April. After the prey is killed it is equally distributed among all the families in the village.

D. Kandha (Tribe of Orissa)

Patnaik (1990) reported that *Kandha* (derived from a Telugu word *Konda* = a hill). The Telugu speaking people call them *Kotuvandlu*. Besides Orissa, these people also live in Andhra Pradesh, West Bengal and Madhya Pradesh. Originally they were hill-dwellers.

There are various sections among the Kandhas. Each section is endogamous through originally they hail from the Kandha community.

The Khandas are Agriculturists. The occupational patterns are varied. The primitive groups are mainly shifting cultivators. They carry on cultivation on swiddens including hunting and food gathering.

The primitive sections produce paddy from the uplands, wherever available at the foot of the hills but mainly depend upon minor millets and cereals. They also produce cash crops like, turmeric, ginger, mustard, niger, black gram, arrow root, etc. The *Dongria* section of the Kandha community produce fruits like, banana, pineapple, orange, jack, etc. All these sections rear livestock. They utilize their cattle wealth mainly for feasts and festivals. But the *Dsia Kandhas* utilize their cattle mainly for ploughing the field.

E. Irular (Tribe of Tamil Nadu)

Ravishankar (1994) conducted a study on role of tribal communities in the conservation of plant genetic resources and documented his findings. The salient findings of his study were given below.

Irular (Tamil, *Irul* means black or darkness). The attribution of the name Irular, to these tribal people may be either due to the darkness of their skin or the thick jungle in which they live. They are also called as *Villians* (bowmen).

Irulars are found scattered in the districts of The Nilgiris and Chengalpattu-MGR, North Arcot-Ambedkar, Tanjore, Tiruchirappalli and Tiruvannamalai-Sembuvarayar in the state of Tamil Nadu.

They live in small groups, in huts that are of conical shape and thatched with palmyra (*Borasuss flabellifer*) fronds. Each hamlet has a headman, this headmanship is not hereditary in Irular communities.

They collect honey from the wild by climbing cliffs at night by a ladder made up of twisted creepers. The chief source of livelihood of the womenfolk is by husking paddy in the neighbouring villages. They also gather and sell firewood for subsistence.

Irulars depend on plant kingdom for their day to day living, by collecting edible tubers (yams) namely 'irular khizanghu' (*Dioscorea opoositifoliai*) and 'valli khizanqhu' (*Dioscorea pentaphylla*). These tubers are repeatedly washed and cooked before consumption. They also possess good knowledge in the identification and skill in snake catching. They also have knowledge of herbal medicine used as antidote for snake bite. For example leaves of *Andrographis paniculata* and *Vernonia anthelmintica*, tubers of *Corallocarpus epigaeus*, roots of *Rauvolfia tetraphylla* and the bark of *Cassia fistula* to treat the snake bites. These plants are used either individually or collectively. This knowledge is used by them to treat outsiders also for which they are rewarded by means of money or other kind. These people collect the plant parts used, for example leaf or root or bark, dry it and store it for use whenever it is needed.

The staple food of these people are samai and rice, kambu (*Sorghum* sp.) and thinai (*Setaria* sp.) are also consumed, usually in the form of porridge. Their diet also consists of vegetables and pulses like avarai, thuvarai and uluval or kollu.

In this community there is a tradition of worshipping prehistoric Celts (axe-like instruments), stone implements and images placed in huts within the forests. The reverence towards such sacred groves, have resulted in the preservation of the flora and fauna within such areas.

Their economic activity is agriculture dependent and in cattle rearing and poultry, agriculture being the primary occupation of the community, and traditionally they were subsistence farmers with knowledge in the hunting of wild animals.

Of late due to depletion of forest resources on which they were mostly dependent, these people labourers for wages in the cultivated fields of the so called upper class communities. To some extent with the help of government policies they could acquire some piece of land and go for their own crop cultivation. They concentrate mainly on minor millets and vegetables. The major problem faced by this tribal group is water scarcity as it is the case with other tribal communities also. According to the elderly tribal people the depletion of water resources is mainly due to the degradation of natural vegetation.

F. Kothar (Tribe of Tamil Nadu)

Ravishankar (1994) has given the characters of Kothars that they are named after their settlement in Kothagiri of The Nilgiris. They are renowned people of the Kothagiri region and denoted agriculture as their main occupation. Though their nature of crop production and cropping pattern has been drastically changing in the recent years, they still depend on agriculture as their main occupation. They go for the cultivation of vegetables, tea and plantation crops. They conserve the traditional varieties of minor millets like Varagu, Samai and Thenai. These crops are still their important food crops.

They are good conservers of wild biological diversity also. They keep a portion of their land as sacred groves or *kavus*, to which they attach lot of sanctity. Because of this many plant species and associated fauna have been preserved. They are conserving many tree species like champakam (*Michelia champaka*), vella champakam (*Plumeria alba*), pala (*Alstonia scholaris*), pandam (*Canarium strictum*) and asokam (*Saraca asoka*) which they prefer for shelter, and consider it to be the residence for some of their benevolent spirits. Depending on the availability of the species in their surroundings any one of these will be selected, ad dedicated to the spirit/God.

A great majority of them own at least a small piece of agricultural land. The women are equal to the men in all agricultural activities. They also supplement their income by collecting minor forest produces and by working as labourers with forest contractors and in plantation estates. They also collect minor forest produce such as honey, soap nut (*Sapindus emariginata*), kadukai (gall nut), pongam (*Pongamia glabra*) seeds and ettikai (*Strychnos nux-vomica*). Some of these communities have taken up the practice of apiary and sericulture during the past few years, as additional source of income generation.

The variability in their traditional crops is threatened by the intrusion and rapid spread of high yielding varieties and commercial crops like tea. Many farmers recognize and appreciate the value of growing their native crop varieties. It is pointed out that most of their traditional crops are drought tolerant, pest/disease resistant, nutritious and require no/less external inputs.

G. Paniya (Tribe of Kerala)

Anil Kumar (1995) reported that Paniyan means one who works in the field, and they were supposed to be work for others mainly the non-tribes. This group is the dominant tribals in the State. About 72 per cent of the total Paniya population of the State are found in Wayanad district.

He added that they worship trees, banyan tree (*Ficus bengalensis*) is important for them. They will not cut such trees and believe if any one attempts, for that, they face serious problems or fall sick. They consume wild roots, edible and medicinal herbs, fish, prawns, crabs which are usually found in the wilderness.

H. Kurichia (Tribe of Kerala)

Anil Kumar (1995) reported that *Kurichias* are the "high class" people among the hill tribes, occupying the highest social and economic position among them. They are basically Agriculturists having their own lands. They preserve their habitats in almost pristine conditions even now and conserved many of the traditional varieties of different crops.

They once practiced *punam* (shifted cultivation) largely for *muthari* (ragi) and for a type of hill rice. Usually women were employed in shifting cultivation while men go for hunting. Presently they are engaged only in wet land cultivation, a minimum of six varieties of paddies has been conserved because of their efforts in the traditional agricultural practices. They live in joint family system and work together in the field. This system has helped to a great extent in preserving their paddy lands whereas many among the settlers are giving up paddy cultivation because of higher labour charge and other reasons.

I. Malayali (Tribe of Tamil Nadu)

Ravishankar and Selvam (1996) reported about Malayali tribes that Malayali tribes were hill dwellers, hill people, inhabitants of the hills or rulers of the hills.

Life style of a majority of Malayalis, at least until last twenty years or so has been dependent on the utilization of various natural resources including plant species. They practiced subsistence economy and had been almost self sufficient. Recognizing the value of the local plant resources they have developed efficient management practices. Today, dependence on various plant genetic resources and the knowledge acquired over years of association with these resources is on the brink of total disappearance. Process of modernization and outside contacts have largely been responsible for changing life styles of these people. Pressures of cash economy and growing population have been major contributing factors for ecological degradation. Subsistence mode of agriculture has partially or fully been replaced by cash economy. Their agro forestry system that promoted growing of trees in and around the agricultural lands has disappeared to a large extent.

Agriculture of The Nilgiris is characterized by high degree of agro-biodiversity. It has various traditional cultivars, which not only have significance in terms of agriculture, but also have socio-cultural importance. The introduction of tea and other cash crops and slow decline in the traditional crops such as minor millets and the disappearances of certain land races have bearings on the agro-biodiversity of The Nilgiris.

From the above literature it is considered that the traditional way of crop production has been adopted by various tribal groups by one way or the other. Among the discussed the Kothar and the Irular tribal groups has been taken for the study. They are the two important tribal groups of The Nilgiris having their main occupation as agriculture.

2.4. Participation of Tribal Women in Agro-biodiversity Conservation and Relationship with the Profile Characteristics

Participation of tribal women in agro-biodiversity conservation and relationship with the profile characteristics has been explained in the following sub-heads:

Participation of Tribal Women in Agro-biodiversity Conservation

Division of labour between men and women varied from region to region according to social custom and community. Women's participation in agriculture was found to be highest among the tribes followed by the Scheduled Castes. The degree of participation declined with higher position in the social hierarchy as opined by Majumdar (1975).

Johnson (1983) identified six levels of participation in extension activities *viz.*, (i) participation in response to an order or to force, (ii) voluntary participation simulated by reward (iii) voluntary participation prompted by awareness, (iv) participation by giving suggestions and making criticism aimed at the improvement of an activity, (v) participation by taking initiative and (vi) participating by creativity.

Sharma and Prasad (1983) observed that the tribal women had to work harder than men. They cared the household and children; help the men folk in agriculture etc. Besides, they had to collect firewood from the forest often and it became necessary for them to work for wages. They were less paid then men folk.

Bhuyan and Tripathy (1988) observed that in tribal areas, women worked with men in sowing, application of fertilisers and pesticides. The tribal women were taking the arduous work of cultivation in contrast to the women in plain or coast. Nearly 90.00 per cent of the agricultural operations were shared by the women.

Deshpande (1988) stated that tribal women shared 75.00 per cent of all the farm work. The daughters-in-law performed 4.90 per cent of it. In dairy, poultry and goat rearing, tribal women carried out 68.00 per cent of the work. Cleaning and feeding of cattle and the entire care of poultry was in their incharge.

Tribal farm women participated in all the activities *viz.*, forking, clod breaking, formation of ridges and furrows, formation of drainage channel and application of FYM, connected to main field preparation and also the weeding, harvesting except marketing as opined by Manjula (1991).

Singh and Sharma (1993) found that hill women were mostly involved in repetitive and monotonous operation like transplanting, harvesting and threshing. They also found that 58.38 per cent of the farms were managed by men, 16.59 per cent were managed by women and 24.88 per cent were jointly managed.

Patel and Sandagi (1995) stated that majority of the farm women participated in harvesting (97.17 per cent), planting (91.51 per cent), weeding (88.68 per cent), feeding the animal and poultry birds (88.19 per cent), post harvest operation (74.47 per cent), nursery raising and planting of vegetables (74.47 per cent), livestock care (75.47 per cent), preparation of FYM and compost (63.21 per cent) and winnowing (54.72 per cent).

Parvathi (1995) opined that majority of the women agricultural labourers were involved in transplanting (80 per cent), transplanting and spreading of seedling (60 per cent) in paddy, reaping the crop (98 per cent) and bundling and transportation (94 per cent).

Chakrabarty (1997) indicated that while ploughing was the monopoly of men in wet cultivation, the tribal women's work load nearby no means less than their men. In sowing, uprooting the seedlings, transplantation, weeding, harvesting and threshing both tribal men and women were participating. But uprooting the seedlings, transplanting and weeding were predominantly done by women.

Sharma (1997) revealed that all vegetable crops were cultivated by women and maintained by them, the other crops were cultivated jointly by men and women but post production aspects of agriculture were taken care of by women. Thus women played an important role on separation of seed material from the grain after harvesting, dehusking and milling of grains and storing of seeds. He has also suggested that among the tribal communities of this area men and women take equal parts in all agricultural operations. Ploughing, digging etc., are usually taken up by the male members and time consuming work such as weeding, transplanting are taken up by women. The present day land races of all field crops have been developed by mass selection of single ear heads.

Marimuthu (1998) found that the entire responsibility of raising and nursing the vegetable garden except land preparation is a women's job. Along with vegetables, the women also grow pulses like padinettu maniyan (*Lablab purpureus* L. sweet var. purpureus), Avara payar (*Lablab purpureus*). Different types of tuber crops are grown for household consumption. Chembu (*Colocassia esculenta*), chena (*Amorphophallus campanulatus*), and katchil (*Dioscorea* spp.) are the major crops cultivated by this people. Petiole of different varieties of chembu like para chembu, neela chembu and cheru chembu (both in cultivated and wild) is used as greens.

Jamatia (1999) found that the tribal women were found to participate in the activities like stubble collection, forming ridges and furrows, forming drainage and application of FYM, connected to main field preparation in various degrees self doing, assisting and supervising. He added that majority of the tribal women had participated in the sowing, weeding, harvesting and post-harvest activities by self doing. He has also suggested that the decision to grow certain paddy varieties and the area under the cultivation of these varieties is taken based on various factors like nutritive, palatability and size of the grains by both men and women. The choice of the vegetables to be grown is generally taken by women. The women's concern for providing the family with nutritive vegetables is great that sometimes she has to request and convince her men to grow few vegetables along with the cash crops like ginger which is generally for the market. It is the women's idea to grow vegetables along with the ginger. The men are just worried about the income; the women are concerned about providing the family with food. Another activity which women and sometimes children take interest is the gathering of *kumilu* (mushroom) from cultivated fields and periphery of forests.

Bhuvaneshwari (2005) revealed that majority of the tribal women possessed high level (61.00 per cent) of participation in agro-biodiversity conservation. She also observed that there was medium participation (69.00 per cent) in *in situ* conservation and high (89.00 per cent) level of participation in *ex situ* conservation.

Relationship of Profile Characteristics with the Participation

Achanta (1982) revealed that patterns of division of labour between men and women varied from region to region according to social customs. Cultural norms affected women's participation in agriculture.

Santha Govind (1984) found that the characteristics namely extent of land, social participation and annual income had shown significant but negative association with the extent of participation in farm activities, while the rest of the characteristics namely age, caste, education, occupation, farm experience, knowledge level, type of family, size of family, mass media exposure and extension agency contact showed non-significant relationship.

Manjula (1991) observed that significant correlation between extent of participation of tribals in agriculture and allied activities with age, educational status, family size, farm size and livestock possession existed and the intensity of the relationship had been maximum with family size, educational status, age, livestock possession and farm size. This clearly indicated that the above variables had direct relationship with the extent of participation bringing out that influence on the extent of participation had been remarkable.

Seema and Prasad (1991) stated that four variables *viz.*, age, educational status, contact with extension agency and family educational status were found to have significant predictive function to explain the variation in role performance of farm women.

Sujatha (1996) found that farming experience or mass media participation showed positive and significant association with role performance of farm women in agriculture and allied activities. Social participation, job preference and knowledge level showed negative and significant association with role performance of farm women in agriculture and allied activities.

Premavathi (1997) stated that eighteen characteristics like age, farming experience, experience in home activities, family status, social participation, farm power status, material status, mass media exposure, cosmopoliteness, marketing behaviour, scientific orientation and socio-economic status had positively significant association with their involvement of rural women in farm activities but other variables namely educational status, occupational status, farm status, credit orientation, innovativeness and change agency contact had negatively significant association with their involvement of rural women in farm activities.

There was positive correlation between size of land ownership and economic participation of tribal women but found negative correlation between educational status and economic participation of tribal women as concluded by Chakrabarty (1997).

Jamatia (1999) revealed that the characteristics namely occupational status, fatalism-scientism, religious belief and gregariousness showed positive and significant association with participation of tribal women in agriculture and allied activities, whereas educational status, annual income, mass media participation, value orientation, conservatism-liberalism, credit orientation, economic motivation, progressivism-traditionalism and achievement motivation showed negative and significant association with participation in agriculture and allied activities. In addition, he expressed that occupational status had contributed much for participation in agriculture and allied activities.

Bhuvaneshwari (2005) revealed that age, educational status, experience, social participation, innovativeness and economic motivation showed positive relationship with participation in agro-biodiversity conservation and experience, social participation, innovativeness and economic motivation showed positive influence on the participation in agro-biodiversity conservation of the tribal women.

2.5. Human Dependency on Agro-biodiversity

Swaminathan and Jana (1992) stated that agro-biodiversity contains (1) genetic resources for food and agriculture (harvested crop varieties, livestock breeds, fish species and non-domesticated wild relatives, forests, range lands and aquatic ecosystems); (2) components of agriculture biodiversity that provide services (non harvested species within production ecosystems that support food provision, including pollinators, nutrient cycling species, natural enemies), as well as services important to agriculture such as water catchment, maintaining water quality, carbon sequestration, and maintenance of local wildlife and habitats; and (3) biotic factors affecting the agricultural biodiversity and the socio economic factors that are shaping it through human activities and management.

Bunning and Hill (1996) stated that agro-biodiversity encompasses many types of biological resources related to agriculture, namely, (i) Genetic resources – the essential living material of plants and animals, (ii) Edible plants and crops, including traditional varieties, cultivars, hybrids and other genetic material developed by breeders, (iii) Livestock (small and large, lineal breeds), (iv) Freshwater fish, (v) Soil organisms vital to soil fertility, structure, quality, and soil health, (vi) Naturally occurring insects, bacteria and fungi that control insect pests and diseases of domesticated plants and animals, (vii) Agro ecosystem components and types (polycultural/monocultural, small/large scale, rainfed/irrigated, etc.) indispensable for nutrient cycling, stability and productivity, and (viii) "Wild" resources (species and elements) of natural habitats and landscapes that can provide services such as, pest control or ecosystem stability, to agriculture.

FAO (1999) had reported that agro-biodiversity account for or influence the following, (i) Increased productivity, food security and economic returns through diversified products and income opportunities, (ii) Providing sources of improved human nutrition and medicines, (iii) Reduced pressure of agriculture on fragile areas, forests, and endangered species, (iv) Contributing to sustainable intensification, (v) Reduced dependence on external inputs, (vi) Making farming systems more stable, robust and sustainable, (vii) Contributing to sound insect pest and disease management, (viii) Conserving soil, (ix) Increased natural soil fertility and health and (x) Conserving ecosystem structure and stability of species diversity.

Pandey (2004) stated that agro-biodiversity occupies a unique place within the biological diversity and is essential to satisfy basic human needs for food and livelihood security. Agro-biodiversity is actively managed by the farmer.

Trivedi (2006) reported that agricultural biodiversity is not only the result of human activity but human life is dependent on it not just for the immediate provision of food and other natural resources based goods, but for the maintenance of areas of land and waters that will sustain production and maintain agro ecosystems and the wider biological and environmental services (biosphere). He also suggested that agricultural biodiversity provided:

- ✰ Sustainable production of food and other agricultural products emphasising both strengthening sustainability in production systems at all levels of intensity and improving the conservation, sustainable use and enhancement of the diversity of all genetic resources for food and agriculture, especially plant and animal genetic resources, in all types of production systems.
- ✰ Biological or life support to production emphasising conservation, sustainable use and enhancement of the biological resources that support sustainable production systems, particularly soil biota, pollinators and predators.

Ecological and social services provided by agro ecosystems such as landscape and wildlife protection, soil protection and health (fertility, structure and function), water cycle and water quality, air quality, carbon-dioxide sequestration, etc.

Importance of Agro-biodiversity

According to Singh and Sharma (1993) the importance of the gap between the market price of environmental resources and their value to individuals and societies is only gradually coming to be understood. The causes of the gap are much better understood. One of these was ignorance or uncertainty about the social consequences of private actions. Secondly, the structure of right that encourages people to ignore the known social consequences of their actions. In a country like India with the increasing human and livestock population, concrete actions and sound strategies are easy to formulate but difficult to implement. Human beings have a tendency to prefer present goods to future goods and in the process, they overlook the side effects.

Ramachandran and Nagarathna (2003) has reported that biodiversity is important to humankind in fulfilling its needs by way of providing food (80,000 species), medicine (20,000 species), drug formulations (8000 species) and raw materials (90 per cent from forests) for industry. India is one of the mega biodiversity countries in the world (hosting 75,000 species of fauna and 45,000 species of flora). India is also a vavilovian centre of origin and diversification of cultivated plants with approximately 167 species of agri-horticultural crops and 320 species of their wild relatives. In this regard, India is identified as one of the important biodiversity pools, known for genetic ecological and economic prudence.

Kumary (2008) reported that biological diversity or biodiversity, a recently coined term, denotes species richness (of plants, animals and micro-organisms) occurring as an interacting system in a given habitat. She also noted that according to the U.S. Office of Technology Assessment, it is "the variety and variability among living organisms and the ecological complexes in which they occur". It generally refers to the variation among living organisms, *i.e.*, from molecular unit to the individual organism and then onto the population, community, ecosystem, landscape and biospheres levels, referred to also as genetic diversity, species diversity, ecosystem diversity. Biodiversity is thus the assemblage of different life forms. Biodiversity studies include: (i) a systematic examination of the full array of organisms on this globe, and (ii) a study of the methods by which diversity can be maintained and used for the benefit of mankind. Biodiversity cannot be replaced because the species becomes adapted in a given habitat after a long course of evolution.

Social Dimensions of Agro-biodiversity and Sustainable Livelihood

FAO (2006) mentioned the social dimensions of agro-biodiversity and sustainable livelihood was as follows:

Enhancing Appropriate Coping Mechanisms

Agro-biodiversity and indigenous knowledge are relevant ingredients in local coping mechanisms to food insecurity, since they represent the locally available resource base for agriculture, food production, and healthcare. This grassroots dynamic demands adequate recognition and support from agricultural policies and programmes.

Exploring Agro-biodiversity-based Livelihood Options

Agro-biodiversity encompasses a number of options for livelihood and economic development in rural communities, but they are severely constrained in the existing socio-economic framework. The effective use of agro-biodiversity to enhance rural economies requires active support at all levels, including policies, institutional mechanisms, and targeted interventions; for instance in research projects, extension programmes, rural micro credit schemes, and market policies.

Gender Equity

Gender inequalities aggravate food insecurity impact among rural women, whilst constraining their roles and capacity to combat such constraints at household and community levels. Rural women play numerous and vital roles throughout the agriculture-food-nutrition-health complex, including particularly the conservation and use of agro-biodiversity, as well as the transmission of indigenous knowledge.

Community Mobilisation

In some rural areas is likely to undermine community dynamics, which are however, increasingly important to support affected households, mitigate the impacts, and launch rural development. It may erode the community-based conservation of agro-biodiversity, as well as social reproduction systems that facilitate the exchange and transmission of indigenous knowledge. In this context, community-led initiatives that promote the conservation, use, and management of biodiversity and the associated indigenous knowledge are highly important. Such community mobilisations are relevant to enhance local capacity to cope with food insecurity. They are equally important to maintain biodiversity and indigenous knowledge as grassroots components for rural development. In particular, a rural emergency situation, the active stance for conserving and making available agro-biodiversity and indigenous knowledge becomes increasingly important to maintaining the prospects of reconstruction, development and autonomy among surviving community members, including the youngest generations.

2.6. Women and Environmental Issues

Warren (1991) observed that public opinion is one of the most important areas of work for promoting environmental concerns and interests. Women in local organizations have a first hand knowledge of the impact of environmental degradation in their communities. Women can help publicize regional examples of the abuse of the environment in their communities so as to raise awareness that can lead to more effective political action. There is growing evidence that women in several countries around the world are taking central roles in the grassroots environmental movement and their is increasing belief that development policies that do not involve women and men alike will not be successful in the long run.

Chelala (2002) observed that women are traditional protectors of the environment. A world survey on public attitudes toward the environment sponsored by the United Nations Environment Programme showed that women, when compared to men, are more likely to choose a lower standard of living with fewer health risks over a higher standard of living with more health risks. The platform for action adopted by the Fourth World conference on Women in Beijing identified the need to involve women more actively in environmental decision making at all levels, as well as to incorporate their perspective in all strategies leading to sustainable development.

Anonymous (2002) of the Centre for symbiosis of Technology stated that women's participation in the formulation, planning and execution of environmental policy continues to be low. At the same time, the international community had recognized that without women's full participation, sustainable development cannot be achieved. Women have a key role to play in preserving the environment and natural resources, and in promoting sustainable development. For example, women still have the main responsibility for meeting household needs and are therefore a major force in determining consumption trends. As such, women have an essential role to play in the development of sustainable and ecologically sound consumption and production patterns.

Anonymous (2003) of the Centre for Symbiosis of Technology listed the steps taken by various Governments around the world to promote sustainable development and to incorporate a gender-perspective into their national environmental policies and programmes.

- ☆ Canada has promoted the mainstreaming of a gender-perspective in the sustainable management of freshwater, oceans and forests, the protection of biodiversity and combating desertification.
- ☆ The National Directorate for Equality of Women in Colombia is working with the Ministry of the Environment to incorporate a gender-perspective in the planning and implementation of its programmes.
- ☆ Portugal has appointed a woman as the Minister of the Environment.
- ☆ In Tunisia, women hold 19 per cent of senior management positions in the Ministry of Environment and Regional Development. Women constitute 36 per cent of Ministry's staff.
- ☆ In Jamaica, women make up 37 per cent of the technical staff of the forestry department. Changes in recruitment strategies have raised the percentage of women working as administrative staff in the environmental sector to 69 per cent.
- ☆ In China, as on 1997, 38 per cent of the total staff working in Environmental Protection Department was women.

Sehgal (2004) observed that fuel, fodder and water collection, are the accepted responsibilities of women. As the environment degrades, these basic necessities become difficult to collect. The time a woman spends on gathering fuel, fodder and water, as well as attending to household work, agricultural work and animal care, reduces her efficiency. In the hilly areas, the time devoted to these activities far exceeds the time spent on the same activities in the comparatively developed state of Kerala, where land reforms have resulted in providing relief in securing fuel requirements. The following were the reforms:

- ☆ Traditionally, women have dealt with non-monetised biomass based subsistence economy of the household *i.e.* firewood, cowdung, crop wastes, organic manure etc. In comparison, men and tend to destroy nature to earn cash, even if its means creating hardship in their own families for their womensfolk to collect fuel and fodder, *e.g.* Sale of herbs and wood. The upshot is that women work as unpaid labourers on family farms with a greater role that man in operation decision-making.
- ☆ The population pressure has increased male migration, which in turn adds to be women's workload. In effect, this means that women's responsibilities extend from the household duties to working in the field as well. As the time required for fuel and fodder collection grows and firewood becomes scarce, cow-dung previously spread on the fields, is used in the kitchen, thereby depleting soil resources and causing a negative effect on the livelihood of local people and the environment. The absence of the male family member(s) adversely affects production. In Haryana, the agricultural workload of women increased after men took up jobs in cities.
- ☆ They are active workers in afforestation, principally, because they are the primary victims of deforestation, especially when deforestation for settlement purposes has taken place. They are involved in mobilizing and

making selective use of forests and other natural resources. For *e.g.* They apply manure to the soil and help regeneration of the forest by regulating its use. They are aware of the forest land, livestock and human resource chain essential for sustainable development, *e.g.* The Kerahosahalli village of Uttar Kanara district in Karnataka where the 'Appiko' movement is active. They demand fuel and fodder trees which meet household needs, thus lending a practical aspect to their daily requirements *e.g.* In Chamoli and in the Thar Desert.

- ☆ Skills are handed down from mothers to daughters to help strengthen the bond between man and environment. This is sustainable development. Women mobilize, conserve and preserve recourses in their roles as mothers, wives, daughters and farmers.

Sometimes, development projects contribute to the damage of our natural and culture environment, whereby women are the primary victims. For *e.g.* Dam constructions, quarrying. Given the vital link of women in the development process, they have necessarily to be equal partners in the decision making and implementation process. The foremost concern of a society should be to recognize gender specific problems, and to accept the factors which ensure benefit to women.

It can be concluded that the active participation of women in think tanks and in environmental training activities is allowing them to educate both the public and policymakers about the critical links between women, natural resources and their use, and sustainable development. In that regard, women have better access to local environmental issues and how to approach them than men. Women have often had a leadership role in reducing unnecessary use of resources, promoting an environmental ethic and recycling resources to minimize waste. In some regions, women are generally the most stable members of the community, since men frequently pursue work in other regions or countries. In those conditions, women are in a better position to safeguard the environment and ensure an adequate and sustainable allocation of resources leading to better and sustainable livelihood.

2.7. Constraints and Suggestions of Tribal Women

Mahapatra (1978) opined that tribal people attributed lack of adequate communication facilities as one main reason for agricultural backwardness of tribal villages. The diffusion of knowledge of improved agriculture was limited by the communication gap in the tribal society.

Oberoi and Moorthi (1986) reported that the Gabbi tribal people of Himachal Pradesh continued to follow old methods of food production and sheep rearing. Subsistence farming is practiced largely because of the difficult hilly terrain, technological backwardness and poor means of communication.

Kokate (1988) reported that traditional beliefs held by the tribals were bottlenecks in the adoption of improved practices related to agriculture and cattle feeding.

Gnanadeepa (1991) reported that 76.00 per cent of farmers expressed the availability of labour and wages were major constraints for adoption of indigenous practices.

Manjula (1991) indicated that the problems of tribal families in agriculture were low price and fluctuation of price for farm produce (94.00 per cent), high cost of inputs (88.00 per cent), inadequate and high wage rate for labourers during peak season (86.00 per cent), difficulty in loan repayment (54.00 per cent), paucity of credit facilities (48.00 per cent), lack of transport facilities (28.00 per cent) and non-availability and insufficiency of inputs in time (20.00 per cent).

Rajulashanthy (1991) revealed that the burden of work in the farm and at home and lack of knowledge in health care in livestock was reported as problems by more than 75.00 per cent of farm women. Sixty to seventy per cent of the farm women reported lack of knowledge in health care in crops, lack of marketing facilities, lack of demand in the market and lack of remunerative price as problems experienced. Small size of holdings and lack of facilities for animals and birds were reported as problems experienced by 59.00 and 42.00 per cent of farm women respectively.

Balasubramanian (1992) reported that during the implementation of Lab to Land programme among the tribes, they faced the following problems in accepting new technologies *viz.*, unfavourable weather conditions, insufficiency of inputs supplied, lack of transport facilities, exploitation by village leaders and soil erosion.

Nandini (1995) indicated that the reasons accounted for the non-adoption of contour cultivation were small size of holding, lack of knowledge, difficulty in adoption and the fact that the particular practice was not needed. For the non-adoption of contour bunding, the reasons attributed were lack of financial facilities, lack of knowledge, non-co-operation from fellow farmers, water stagnation near the bunds and fragmented size of holding.

Ramu (1997) observed the constraints for tribal development during 1995 with relative importance were lack of credit institution, poor infrastructure facility, lack of communication facility, inadequate marketing, lack of storage and processing facility, non-availability of credit, lack of technical guidance, low agricultural production, poor technology, problematic land conditions, insufficient input, low family income, isolation and exposure, poor quality livestock, lack of transport facility, poor housing facility, inadequate school facility, inadequate health facility, shortage of drinking water and ill-health.

Lakshmana (2000) observed that most of the farmers expressed high labour requirement, more time requirement, bulkiness of indigenous inputs and climatic conditions as constraints in adoption of Intergrated Farming Practices (IFP). Around half the respondents had expressed treating backward; lack of interest in IFP and less effectiveness were acted as barriers in adoption of IFP. Considerable percentage of farmers expressed lack of proper plan, competition from modern technologies, and lack of sufficient knowledge and non-availability of skilled persons as constraints for adoption of IFP.

Selvarani (2000) indicated that among the various physical constraints of tribal farmers studied, remoteness of their farm was expressed as the major physical constraint, followed by labour problem and poor quality of inputs available in the market. Lack of training and lack of technical guidance and non-availability of extension service were expressed as the major extension constraints. Major economic constraints encountered by the respondents were high cost of inputs, high wages to labourers and high rate of interest. Incidence of pest and diseases, high risk and uncertainty of returns, undesirable climatic factor, lack of post-harvest technology and lack of plant protection methods were the major technological constraints. High transport cost and lack of transport facilities were the major marketing constraints felt by the tribal farmers.

Bhuvaneshwari (2005) observed that complexity of practice, lack of training, unfavourable climatic factors, requirement of physical strength, lack of credit and lack of marketing experience as the major constraints faced by the tribal women. Further she revealed that lack of communication facilities, unfavourable weather conditions, labour problem, inadequate rainfall, technological backwardness, traditional beliefs, price fluctuation, remoteness of the farm were the major constraints of the tribal women.

2.8. Environmental Education: Approaches and Dimensions

Stapp (1982) recognized environmental education as a process aimed at developing a world population that is aware of and concerned about the total environment and its associated problems, which has knowledge, attitudes, motivations, commitments and skills to work individually and collectively toward solutions of current problems and the prevention of new ones.

UNESCO (1990) reported that the environmental education is the process of recognizing values and clarifying concepts in order to develop skills and attitudes necessary to understand and appreciate the inter relatedness among man, his culture and his bio physical surroundings. Environmental education also entails practice in decision-making and concerning environmental quality.

Tomar (2007) opined that environmental education is a process to promote the awareness and understanding of the environment, its relationship with man and his activities. Environmental education is both a style and subject matter of education. Style means the approach and subject matter means content.

Goals and Guiding Principles of Environmental Education

Following are the goals of environmental education as identified by Tomar (2007):

- To foster clear awareness of, and concern about, economic, social, political and ecological interdependence in urban and rural areas.
- To provide every person with opportunities to acquire knowledge, values, attitudes, commitments and skills needed to protect and improve the environment.
- To create new patterns of behaviour of individuals groups and society as a whole towards the environment.

He also identified the guiding principles of environmental education as follows:

- Consider the environment in its totality-natural technological and social, (economic, political, technological, cultural, historical, moral, and aesthetic)

- ☆ Be a continuous lifelong process, continuing through all formal and non formal stages.
- ☆ Be interdisciplinary in approach, drawing on the specific content of each discipline in making possible a holistic and balanced perspective.
- ☆ Examine major environmental issues from local, national, regional and international point of view so that students receive insights into environmental conditions in other regional areas.
- ☆ Focus on current and potential environmental situations, while taking into account the historical perspective.
- ☆ Promote the value and necessity of local, national and international cooperation in the prevention and solution of environmental problems.
- ☆ Explicitly consider environmental aspects in plan for development and growth.
- ☆ Enable learners to have a role in planning their learning experiences and provide an opportunity for making-decisions and accepting their consequences.
- ☆ Relate environmental sensitivity, knowledge, problem solving skills and value clarification to every age, but with special emphasis to the learners own community in early years.
- ☆ Help learners discover the symptoms and real causes of environmental problems.
- ☆ Emphasize the complexity of environmental problems and thus need to develop critical thinking and problem solving skills.
- ☆ Utilize diverse learning environments and a broad array of educational approaches to teaching learning about and from the environment with due stress on practical activities and first hand experience.

Curriculum for Environmental Education

Sharma (1997) has suggested that curriculum is all purposeful activities, planned inside or outside the class and earlier it was considered as a single subject. But now it belongs to the curricula of all subjects.

Further he suggested the following stages for curriculum development:

- ☆ Operationalizing environmental education general goals into intermediate sub goals and objectives
- ☆ Delimiting curriculam scope (extent of coverage in terms of objective, cognitive, affective and psychomotor) and sequence (planning elements grade wise)
- ☆ Selection and defining contents
- ☆ Identifying teaching strategies and instructional activities
- ☆ Keeping in mind the local, regional and national needs and three considerations *i.e.* socio-economic considerations, content considerations and the pedagogic considerations

- ✰ It should be related to immediate environment of the people. Content and methodology have to be developed to realize the set goals and objectives
- ✰ The four goal levels are ecological foundation level, conceptual level, investigation and evaluation level, environmental action and skill level.

2.9. Indices Framed on Developmental Issues

Venkataramaiah (1983) while studying the impact of High Yielding Varieties programmes on the dissemination of scientific agricultural information among the farmers of Ganjam district, developed a socio-economic Status scale comprising of the following items: occupation, education, caste, family size, family type, children's education, membership in formal organization, outside contact, reading newspapers, size of land holding, animal net farm income, bullock power, agricultural implements, material possession and house structure. Arbitrary weightages were given to the items; one hundred farers were rated on this socio-economic status scale. According to the ranges of the scores, farmers were classified into five socio-economic status categories.

Mansingh (1993) framed another socio-economic status scale incorporating eight variables. First the list of socio-economic status indicator was prepared. Selection on items was done based on the criteria and judges opinion. The agreement among the judges in giving the scores was determined by working out Kendall co-efficient on concordance and it was tested for significance. The final scale was arrived after clubbing related items and grouping them under common variables. The final scale consisted of the eight variables namely, educational status, family status, occupational status, farm status, social participation status, communication status, farm power status and material status.

"**Index can be operationalised as a numerical scale used to compare variables with one another**."

Ganesan (1989) developed Participation Index by using the following formula. This index represented the participation pattern of the individuals in agricultural development schemes which was used for classifying the respondents and also for working out the relationship with the independent variables.

$$\text{Participation Index (PI)} = \frac{\sum w_i e_i}{\sum w_i e_l} \times 100$$

where,

w_i: Relevancy weightage of each scheme

e_i: Extent of participation score

e_l: Maximum participation score

Theodore (1999) computed composite economic efficiency index and composite ecological efficiency index. The economic efficiency index was calculated by pooling the mean values of all the selected nine economic efficiency indicators *viz.*, production efficiency, net return, cultivated land utilization index, technology use level, low-cost technology use level, employment generation capacity, farm family employment level, self reliant level and self sufficiency level. The ecological efficiency index was calculated by adding the mean value of ecological efficiency indictor such as eco-friendly technology use level, organic recycling level, and low external input use level and soil health.

Arunachalam (2002) proposed the Empowerment Index.

$$EPI = SSI + ESI + DMI + LS$$

where,

EPI: Empowerment Index

SSI: Social Status Index

ESI: Economic Status Index

DMI: Decision Making Index

LS: Legal Safeguards

Palmurugan (2002) developed the following index on empowerment:

a) Perceived Empowerment Level Index

Based on the sum of individual scores of the farm women, perceived empowerment score was obtained and perceived empowerment index was worked out by using the following formula. Based on the perceived empowerment level index score farm women were classified into low, medium and high levels of perceived empowerment.

$$\text{Perceived Empowerment Level} = \frac{\text{Total score obtained by an individual}}{\text{Total possible maximum score}} \times 100$$

b) Empowerment Index

Empowerment index was worked out by using the scale followed by Arunachalam (2002) with some modification. The socio-economic status includes educational status, occupational status, family status, farm power status, material status and social participation status. The scoring procedure suggested by Mansingh (1993) was adopted by summing up the sub-item scores, the socio-economic status was obtained and the farm women were classified as high socio-economic status, upper middle socio-economic status, middle socio-economic status, lower middle socio-economic status and lower socio-economic status.

$$\text{Empowerment Index (EI)} = SESI + DMBI + PEI$$

where,

SESI: Socio-Economic Status Index

DMBI: Decision Making Behaviour Index

PEI: Perceived Empowerment Index

c) Socio-economic Status Index (SESI)

Based on the empowerment index and using the mean and standard deviation the farm women were classified into low, medium and high levels of empowerment

$$\text{Socio-Economic Status Index (SESI)} = \frac{\text{Total score obtained}}{\text{Total possible score}} \times 100$$

d) Employment Potential Index

The employment potential in the study refers to the opportunities for employment available to farm women in farming. The employment potential for farm women were arrived in terms of mandays/hectare with respect to individual crops. Employment potential was calculated by using the employment potential index.

$$\text{Employment Potential Index} = \frac{\begin{array}{c}\text{Employment opportunities available}\\ \text{for farm women in man days/hectare/crop}\end{array}}{\text{Crop duration}} \times 100$$

Sarojini and Moorthi (2002) developed human development index. This index includes the following three indicators.

1. Fix up the minimum and maximum values for the indicator variables – life expectancy, adult literacy rate, gross enrolment ratio and GDP per capita.
2. Construct separate indices for each indicator variable.
3. Take the simple average of the index for life expectancy, educational attainment index and an adjusted GDP per capita index. The general formula used to construct the index was

$$\text{Human Development Index} = \frac{\text{Actual value} - \text{Minimum value}}{\text{Maximum value} - \text{Minimum value}} \times 100$$

This formula for converting indicator values into index form was used for all the variables except for the income variable.

Sakunthalai (2004) has developed the women empowerment index with the following sub components: decision making behaviour, social participation, economic independence, market intelligence, political participation, extension participation, information seeking behaviour, practicing science based technologies, leadership quality, awards won and maintenance of health status.

From the above review the basic idea of index formulation was studied and the index for sustainable livelihood through agro-biodiversity conservation was developed and the index for participation pattern of tribal women in agro-biodiversity conservation was calculated by adopting the scale developed by Ganesan (1989) and Sakunthalai (2004).

Chapter 3

Research Methodology

This chapter deals with the research techniques and procedures adopted in the study and presented under the following sub-heads.

3.1. Locale of research

3.2. Research design

3.3. Research variables

3.4. Construction of Sustainable Livelihood Index (SLI)

3.5. Techniques of data collection

3.6. Statistical tools used

3.1. Locale of Research

Tamil Nadu is one of the major states of the Southern Zone in India which consists of important tribal groups. The selection of district, block and villages are presented along with the description of study area.

Selection of District

The Nilgiris is one of the "Biodiversity Hotspots" as declared by the National Biodiversity Authority in the year 1998. The green carpet of crops and natural settings always speaks about the various biodiversity of the district provides ample scope for the study. The percentage of tribal population was highest in The Nilgiris district than other districts of Tamil Nadu. (Appendix I). Hence The Nilgiris district was purposively choosen for the study.

Selection of Tribal Groups

There are six tribal groups *viz.*, Dodhas, Irulas, Kurumbas, Paniyas, Kothar and Kattunaickar found in the district. As the study is more related to agro-biodiversity continuous performance in agriculture related activities among the respondents were needed. The Irulars and Kothars depend on agriculture profession for their livelihood. Hence these two groups were purposively selected for the study.

Selection of Block

The Nilgiris District comprised of four blocks. *viz.*, Udhagamandalam, Coonoor, Kothagiri and Gudalur. The Irulars and Kothars were found mostly in Kothagiri block. Hence, the Kothagiri block was selected purposively.

Selection of Villages

Kothagiri, and Kunjappannai villages were purposively selected, considering the maximum number of identified tribal habitations. (Appendix II).

Selection of Respondents

The list of tribal farm women from the selected two villages was obtained from the Horticulture department. A sample size of 200 tribal farm women was fixed for the study. Proportionate sampling technique was followed for the selection of respondents.

Table 3.1. Selection of Respondents

Village	*Settlement*	*No. of Respondents*
Kothar (Pudukothagiri)	I	70
	II	30
Irular (Kunjapanai)	I	40
	II	60
Total		200

Description of the Study Area

In order to develop a better perception, the details about the study area have been highlighted below.

A. Location of the District

The Nilgiris district has many attractions of its own known as "The Queen of hill stations." The Nilgiris district is situated at an elevation of 900 to 2636 meters above MSL. The name "Nilgiris" means blue hills (*Neelam*-blue and *giri*-hill or mountain). There is a belief that people living in the plains at the foot of the hills, should have given the name, The Nilgiris, in view of the violet blossoms of *Kurinji* (*Strobilanthu* sp.) flower enveloping the hill ranges periodically. The district map of The Nilgiris depicted with the study area has been given in Figure 3.1.

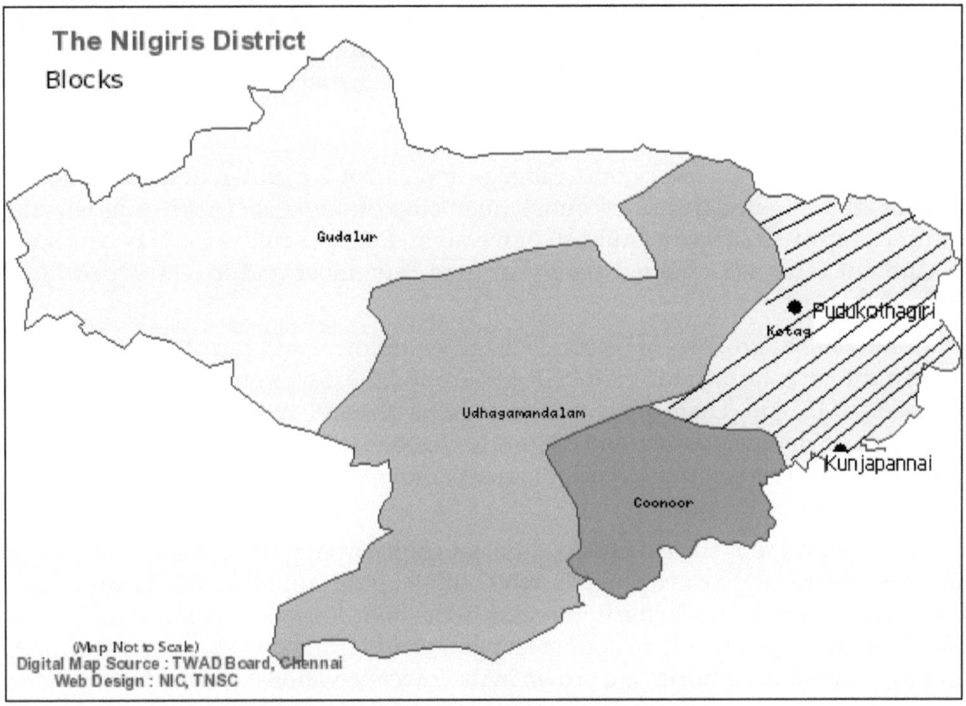

Figure 3.1. Map Depicting the Study Area.

This district marks the western extremity of Tamil Nadu situated in the middle of the western border of Tamil Nadu. Its boundaries are the Coimbatore district in the east and the Kerala state in the south and in the west, and the Karnataka state and Periyar district of Tamil Nadu in the north. The area of The Nilgiris district is 2,549 sq.km.

The wide range of climatic conditions as well as the peculiar topographic features provides a variety of conditions and combinations of temperature, moisture regime, which are faithfully reflected in the diversity of its vegetational communities. The climate is temperate to sub-tropical.

B. Rainfall

The annual rainfall ranges from 950 mm to 1550 mm. The period from the end of November to February is the north east monsoon period with the associated rains ceasing early in January. The summer season is from March to May, June to September is the south west monsoon season. October and November constitute the post monsoon or the retreating monsoon season.

C. Soil

The soils of The Nilgiris district is lateritic in origin and derived from charnockites know as Nilgiris genesis. Climatic conditions favour intense chemical weathering of minerals. The soil is generally deficient in plant nutrients and acidic with pH value ranging from 4.6 to 6.0. It contains large per cent of alumina and iron, which restrict the availability of phosphates to plants. The organic matter content of the soil is high.

D. Cropping Pattern

The annual crops like potato, cabbage and carrot are grown in three different seasons *viz.*, irrigated (February-June), main crop of *Karbogam* (April-August) and autumn crop or *Kadaibogam* (August- November). Paddy is cultivated in two seasons in Gudalur taluk *viz.*, rainfed paddy in June-September and irrigated paddy in December-April.

Commercial growing of beans, cabbage, cauliflower and carrot during normal times has been concentrated mainly in and around the towns of The Nilgiris over an elevation of 1524 m. There are two main seasons generally when the vegetables are grown *viz.*, (1) January to June and (2) June to October. Potato is cultivated throughout the year (*i.e.*) during summer (March-August), autumn (August-December) and winter (January-May).

Temperate crops such as potato, cabbage, cauliflower, plums, pears and peaches are grown in the high plateau of hills, while sub-tropical crops like mandarin orange, tea and coffee are grown in the mid region of the hills. Various humid tropical crops like clove, nutmeg, cinnamon, pepper and ginger and fruits like mangosteen, durian, grapes, rambutan, litchi etc., are grown in the lower elevations. The land use pattern in The Nilgiris district is presented in the Appendix III.

E. Population Details

Total population of the district according to 2001 census was 7,62,141 with a tribal population of 28373 according for 3.72 per cent of the total population. Among the 36 groups of scheduled tribes in Tamil Nadu, Government of India identified six groups as primitive tribal groups, mostly found in The Nilgiris district. The six tribal groups *viz.*, Dodhas, Kothars, Irulas, Kurumbas, Paniyas and Kattunaickar are not evenly distributed in the six taluks.

Paniyas and Mulla Kurumbas live on the lower western slope of the district upto an altitude of 1200 m in Gudalur taluk, whereas Kurumbas and Irulas mainly live in tracts between an elevation of 1200 m and 1500 m on Kothagiri and Coonoor taluks. The Kothas live in places at an elevation of about 1800 m above MSL in Udhagamandalam and Kothagiri taluks, while the Dodhas occupy the top of the plateau at about 5000-7000 feet high in Udhagamandalam taluk.

The Dodhas were once a pastoral community where life revolved around buffalo-herds. Now they have improved a lot and started cultivating crops like potato, carrot, cabbage, beetroot, cauliflower etc. The Kothars, once considered inferior to other tribes due to their eating of buffalo carrion, are progressive group. They cultivate in their own lands. The Kurumbas were once known for their sorcery and witchcraft. Now they work in plantations and are engaged in cultivation. Paniyas were once known for bonded labour, but are now free due to the intervention of the Government.

Among the identified tribal groups Irulars and Kothars were considered to be cultivation groups, hence these two groups were selected for the study.

F. Kothagiri Taluk

Kothagiri has been known for the biodiversity of various agricultural crops, horticultural crops, flowers, fruits and plantation crops.

Kothagiri sub-division comprises of 15 revenue villages covering 14 revenue villages in Kothagiri and Kukkal revenue village in Udhagamandalam taluk. It is surrounded by Erode district in the north and east. Coonoor and Udhagamandalam taluks in the west and Coimbatore district in the south. It is located at an elevation ranging from 400 to 1900 metres above MSL. The average rainfall of the sub-division is 133 mm in rainy days. The maximum temperature ranges from 18°C to 30°C and minimum temperature ranges from 2°C to 14°C. The soil type is predominantly lateritic red loam with pH range of 3.9 to 7.5.

As this sub division is blessed with adequate rainfall both during south west and north east monsoons agro climatic condition is suitable for plantation crops like tea and coffee in large scale. Besides tea and coffee, fruits like pear, mandarin orange and hill vegetables like potato, cabbage, carrot, beans and minor vegetables beetroot, turnip, knol-knol etc., are also cultivated. Thengumarahada village located in plains on the banks of Mayar river forms part of Kothagiri taluk, where crop like paddy, groundnut, soybeans, marigold and banana are being cultivated due to prevalence of tropical climatic conditions. Sub-tropical climate also provides congenial atmosphere for the better performance of spices crops like pepper. Floriculture is also getting momentum at present. There is an encouraging trend in cultivation of cut flowers like carnation, lillium, alstromeria, gladioli, bird of paradise, calalilly etc.

3.2 Research Design

A research design may be regarded as the blue print of those procedures, which are adopted by the researcher for testing the relationship between the dependent variable and the independent variable. Under the given situation, ex-post facto design was selected as an appropriate research design. Ex-post facto design is the empirical investigation in which the investigator draws the inferences regarding the relationship between variables on the basis of such independent variables whose manifestations has already occurred. In this type of research the investigator has no direct control over the independent variables because they occurred much prior to producing their effects. The present investigation deals with a phenomenon, which has already occurred. This design can be used to device studies, to deduce theories, identify behavior phenomenon occurs to make the concepts operational and measurable.

3.3. Research Variables

Participation of tribal women in agro-biodiversity conservation and sustainable livelihood through agro-biodiversity conservation were considered as the dependent variables in the present study. Eleven variables were considered as independent variables to study their relationship and influence on the dependent variables.

Selection of Independent Variables

The independent variables relevant to the study were selected based on the review of literature, discussion with experts and a pilot study conducted in a place similar to the study area. A list of 33 independent variables that may possibly influence the dependent variables was prepared. The experienced extension scientists and farm scientists were requested to evaluate the degree of relevancy of each variable in the list on a three point continuum. The scores of 3, 2 and 1 were assigned for the "More relevant", "Relevant" and "Least relevant" responses respectively (Appendix IV).

Based on the rating by judges, the mean and co-efficient of variation were worked out for all the independent variables. The overall mean and co-efficient of variation were also worked out. The individual variables, which had their mean, score more than the overall mean and it's co-efficient of variation less than the overall co-efficient of variation were selected for this study. Thus, 11 independent variables were selected. The selected variables and their measurement are listed below in Table 3.2.

Table 3.2. Independent and Dependent Variables and their Measurements.

	Variables	*Scoring Procedure*
A.	**Independent variables**	
X_1	Age	Census 2001
X_2	Educational status	Developed by Mansingh (1993)
X_3	Farming experience	Developed by Mansingh (1993)
X_4	Social participation	Developed by Mansingh (1993)
X_5	Information seeking behaviour	Adopted by Jamatia (1999)
X_6	Innovativeness	Developed by Jamatia (1999)
X_7	Economic motivation	Developed by Supe (1969)
X_8.	Progressiveness	Developed by Pareek and Rao (1974)
X_9	Cosmopoliteness - Localiteness	Followed by Jamatia (1999)
X_{10}	Decision making behaviour in agro-biodiversity conservation	Developed by Thenamudha (1996) and slightly modified for the study
X_{11}	Attitude towards agro-biodiversity conservation	Developed by Arunachalam (2002) and slightly modified for the study.
	Perception on agro-biodiversity conservation practices	Developed by Ganesan (1989) and modified for the study
B.	**Dependent variables**	
Y_1	Participation pattern of tribal women in agro-biodiversity conservation	Developed by Ganesan (1989) and modified for the study
Y_2	Sustainable livelihood through agro-biodiversity conservation	Developed for the study

Independent Variables

Age

Age was operationalised as the number of completed years of the respondent at the time of enquiry and the chronological age was taken as a measure. The age has been categorized into three groups namely, young, middle and old as followed in Census (2001). The actual age was taken as individual score.

Category	*Age*
Young	Less than 30 years
Middle	31–45 years
Old	More than 45 years

Educational Status

Educational status was operationalised as the level of literacy possessed by the tribal farm women. The scoring procedure developed by Mansingh (1993) was adopted.

Category	*Score*
a) Illiterate	1
b) Primary education	2
c) Middle education	3
d) Secondary education	4
e) Higher Secondary education	5
f) Collegiate education	6

The maximum and minimum possible score an individual could obtain was 6 and 1 respectively.

Farming Experience

Experience in farming referred to the number of years of experience a respondent possessed exclusively in farming at the time of enquiry. The scoring procedure given by Mansingh (1993) was adopted. 'One' score was assigned for each year of experience. The respondents were further classified into three categories as low, medium and high based on the cumulative frequency method.

Social Participation Status

This referred to the degree of participation of the respondent in formal organizations either as a member or an office bearer. The scoring procedure given by Mansingh (1993) was adopted.

Category	*Score*
Member	1
Office Bearer	2

To classify the respondents under social participation status as low, medium and high, cumulative frequency was used.

Information Seeking Behaviour

It referred to degree of frequency of contact by an individual with various information sources such as institutional, non institutional and mass media were studied. This was measured with the scale used by Jamatia (1999). The total score was computed for each respondent by adding the scores assigned to different response categories.

Frequency	*Score*
Always	3
Occasional	2
Rarely	1

Innovativeness

Innovativeness was operationalised as the perceived intention or readiness of tribal women to adopt new ideas and technologies of agro-biodiversity conservation relatively earlier than other members of the social system. The scoring procedure developed by Jamatia (1999) was adopted in this study.

Sl.No.	*Statement*	*Score*
1.	As soon as it is brought to your knowledge	3
2.	After you have seen other farmers had adopted successfully in their farms	2
3.	You prefer to wait and take your own time	1

The maximum and minimum possible score an individual could obtain was 3 and 1 respectively. The respondents were categorized into low, medium and high based on mean and standard deviation.

Economic Motivation

Economic motivation was operationalised in terms of profit maximization and the relative value placed by a tribal farm woman on economic ends. It was measured with the help of the scale developed by Supe (1969) and followed by Jamatia (1999). The scale consisted of six statements of which first five were positive and the last one was negative. The responses for each statement were rated over a five point continuum;

with strongly agree to strongly disagree. The scoring procedure followed is given below.

Response	*Strongly Agree*	*Agree*	*Undecided*	*Disagree*	*Strongly Disagree*
Scores for positive statement	7	5	4	3	1
Scores for negative statement	1	3	4	5	7

To get final score of economic motivation for each individual, the score of each statement was added. The maximum and minimum score an individual could get was 42 and 6 respectively. The respondents were categorized into low, medium and high based on mean and standard deviation.

Progressiveness

It is the degree to which tribal women were receptive to modern values and practices. For measuring progressiveness of the respondent the scale developed by Pareek and Rao (1974) was adopted. The scale consists of seven statements for which the respondent had to answer 'yes' or 'no'. A positive response was assigned with two score while a negative response received one score. The progressiveness score was obtained by summing up the response of the items. Based on the score obtained, the respondents were classified into low, medium and high category using mean and standard deviation.

Cosmopoliteness-Localiteness

Cosmopoliteness-Localiteness referred to the tendency of an individual to maintain contacts with other social systems (cosmopolite) or to limit his contact within his own social system (localite). The scale developed by Jamatia (1999) was used in this study. The response to each item was measured over a four point continuum. The scale constituted five statements of which, the scoring procedure followed is given below:

Response	*Strongly Agree*	*Agree*	*Disagree*	*Strongly Disagree*
Scores for positive statements	4	3	2	1
Scores for negative statements	1	2	3	4

First three were negative and the last two statements were positive. The scores for the positive and negative statements were summed up to get total score for an individual. The maximum and minimum possible score an individual could obtain was 25 and 5 respectively. The respondents were categorized into low, medium and high based on the mean and standard deviation.

Decision Making Behaviour

The scale developed by Thenamudha (1996) was used. Decision making behaviour was assessed by the use of a five point scale.

Item	*Score*
Self without consulting	5
Consulting the spouse/elders	4
Consulting all the family members	3
Help others in taking decision	2
No participation	1

By summing up all the scores obtained by each item, the decision making behaviour was arrived at for each respondent.

Attitude towards Agro-biodiversity Conservation

Attitude towards agro-biodiversity conservation referred to the opinion of an individual to favour or unfavour the agro-biodiversity conservation. The scale developed by Arunachalam (2002) with slight modification was used for the study. The responses of each item were measured over a three point continuum. The scale consisted of seven statements of which, the scoring procedure followed is given below:

Response	*Highly Favourable*	*Favourable*	*Unfavourable*
Scores for positive statements	3	2	1
Scores for negative statements	1	2	3

There were four positive and three negative statements. The scores for the positive and negative statements were summed up to get total score for an individual. The respondents were categorized into low, medium and high based on the cumulative frequency method.

Perception on Agro-biodiversity Conservation Practices

Perception was operationalised for the present study as the expressed opinion of the respondents about the *in situ* and *ex situ* conservation practices. The scoring procedure developed by Ganesan (1989) was suitably modified for the study. The perception was studied by the means of five selected statements for each *in situ* and *ex situ* conservation practices which were put on three point continuum according to their positive or negative nature. These statements were arrived at by means of discussion with farmers, farm women, field level extension workers, officials and NGOs. The details of the scoring and other related information are presented below.

Perception on *in situ* Conservation

The agro-biodiversity conservation practices that are followed in the field condition were classified as *in situ* conservation measures. The opinion of the respondents on this type of conservation was measured by assigning the score of 3, 2, and 1 and analysed with the following statements.

Sl.No.	*Aspects*	*Levels of Perception*
1.	Maintenance of seeds of traditional crops in the field	Easy/somewhat easy/difficult
2.	Traditional crop cultivation	Easy/somewhat easy/difficult
3.	Livelihood sustainability through *in situ* conservation	Fully supported/somewhat supported/least supported
4.	Improvement in standard of living	Best improved/somewhat improved/least improved
5.	Government policies	Fully supported/somewhat supported/least supported

Perception on *ex situ* Conservation

The off-farm agro-biodiversity conservation measures was measured as the *ex situ* conservation practices. This opinion was based on the difficulty level of the respondents on the perception on the *ex situ* conservation practices.

Sl.No.	*Aspects*	*Levels of Perception*
1.	Development of seed banks	Easy/somewhat easy/difficult
2.	Traditional food habits	Fully supported/somewhat supported/least supported
3.	Cultural values	Fully supported/somewhat supported/least supported
4.	Livelihood sustainability through *ex situ* conservation	Fully supported/somewhat supported/least supported
5.	Government policies	Fully supported/somewhat supported/least supported

Dependent Variables

Two dependent variables were identified for the study. They were participation of tribal women in agro-biodiversity conservation and sustainable livelihood through agro-biodiversity conservation.

Participation of Tribal Women in Agro-biodiversity Conservation

It has been operationalised as the degree to which the tribal women take part in agro-biodiversity conservation through the management and conservation practices. Development of participation index for tribal women in agro-biodiversity conservation has been explained in the following sub heads.

A. Identification of Agro-biodiversity Conservation Practices

The *in situ* and *ex situ* agro-biodiversity conservation practices were gathered from the discussion with the biodiversity experts and perusal of literature. Accordingly the following four components were identified and given in Table 3.

Table 3.3. Agro-biodiversity Conservation Practices.

Sl.No.	*Statements*
I	***In situ* conservation of agro-biodiversity**
	a. Genetic resources
1.	Cultivating traditional crop varieties
2.	Cultivating medicinal plants
3.	Cultivating location specific crops like minor millets
4.	Domestication of wild varieties
5.	Collection of forest produce
6.	Preservation of forest produce
7.	Conserving soil quality and health
8.	Livestock maintenance
9.	Exotic breeds husbandry
	b. Agro-biodiversity systems
1.	Water shed management
2.	Nutrient management for the crops
3.	Organic cultivation
4.	Adopting natural management practices
5.	Indigenous management practices
6.	Using natural pollinators
7.	Utilization of locally available resources
II	*Ex situ* conservation of agro-biodiversity
	c. Biotic components
1.	Maintenance of seed banks of traditional crops
2.	Developing medicinal plant garden
3.	Cultivation of kitchen gardens
4.	Storage of seeds with advanced systems
5.	Preservation of extinct varieties
6.	Preservation of soil biota
7.	Conservation of micro organism population
	d. Social values components
1.	Traditional food habits
2.	Traditional shelters
3.	Indigenous storage practices

Contd...

Table 3.3–*Contd...*

Sl.No.	Statements
4.	Indigenous post harvest methodologies
5.	Indigenous value addition of the produces
6.	Community storage centers
7.	Community crop management practices
8.	Marketing of the produce locally and outside market
9.	Exchange of seeds within and outside the community
10.	Fixing price for the produce

a. Conservation of Genetic Resources

The conservation of genetic resources included the cultivation of domesticated crops, traditional crops, wild varieties, medicinal plants and location specific crops. It also included the collection of forest produces, preserving the forest produces, conserving the soil health and water management practices.

b. Conservation of Agro-biodiversity Systems

The conservation of agro-biodiversity system included watershed management, nutrient management, organic cultivation, natural management, indigenous practices using the natural pollinators and local resources.

c. Conservation of Biotic Components

Maintenance of gene banks, seed banks of traditional crops, gardens, storage of seeds and extinct varieties and conserving the micro organism population were some of the biotic components in *ex situ* conservation.

d. Conservation of Social Values

Cultural and traditional intervention in the indigenous practices, community management methods, exchange of seeds and diversifying agriculture were the identified social values in ***ex situ*** conservation.

B. Measurement of the Participation of Tribal Women in Agro-biodiversity Conservation

The scale developed by Ganesan (1989) to measure the participation was suitably modified and adopted to measure the participation of tribal women in agro-biodiversity conservation.

Relevancy weightage of each practice (W_1, W_2, W_3)

Level of participation score (e_1, e_2, e_3)-low, medium,high: 1, 2, 3 respectively

This was multiplied by the relevancy weightage.

Weightage for participation = $W_1 \times e_i$

The actual participation score = $\Sigma W_i e_i$

i.e. $W_1 e_i + W_2 e_i + \ldots\ldots\ldots\ldots + W_{12} e_i$

The maximum participation score = $\Sigma W_1 e_i$

i.e. $W_1 e_1 + W_2 e_1 + \ldots\ldots\ldots\ldots + We_1$

$$\text{Participation Index (PI)} = \frac{\sum w_i e_i}{\sum w_i e_l} \times 100$$

where,

w_i: Relevancy weightage of each practice

e_i: Extent of participation score

e_1: Maximum participation score

3.4. Construction of Sustainable Livelihood Index (SLI)

Sustainable livelihood through agro-biodiversity conservation was the second dependent variable. Livelihood sustainability was a systematic approach to link the issues of improvement of standard of living, sustainability and development process. The construction of Sustainable Livelihood Index procedure is given below:

Identification and Standardization of Critical Components

Construction of Sustainable Livelihood Index (SLI) for tribal women was attempted after reviewing the various techniques available in the field of scale construction. Guttmann's (1944) scalogram analysis, Thurston's (1947) law of comparative judgments scale, summated rating method by Likert (1949) were analysed. Before construction of the scale, the components were identified.

The components that would reflect the agro-biodiversity conservation and sustainable livelihood were identified through various sources such as perusing literature, contact with the concerned officials of development departments; self help group members, NGO officials and extension personnel. Thus 12 major components were identified for the study.

Selection of Components

The selection of components was done based on the consensus and item analysis approach.

a) Consensus Approach

Consensus approach means a panel of judges evaluated the items chosen for inclusion in the instrument to find out whether they were relevant to the specified area and possess adequate clarity. The list of identified components was distributed to 60 scientists of various institutions. The judges were asked to judge the components on a three point continuum as most relevant, relevant and least relevant. The weightages assigned were 3, 2 and 1 for most relevant, relevant and least relevant respectively.

b) Item Analysis Approach

The Relevancy Co-efficient Index (RCI) for each major indicator was worked out by using the following formula by Ramanna (1999).

$$RC_i = \frac{\text{Total score of all the judges on ith item}}{\text{Maximum score on the continuum}} \times \text{Total number of judges}$$

Thus the RC_i value for 56 indicators was worked out. The RC_i value for the component ranged from a minimum of 0.40 to 0.80. The RC_i value actually indicated the strength of the components considered in reflecting the level of sustainable livelihood. Therefore it was decided to select only those components with RC_i value > 0.75. Thus 12 components were selected. Again indicators were checked with the criteria as proposed by Trivedi (1963), Venkataramaiah (1983), Mansingh (1993) and Theodre (1999) for the selection of items (Appendix V) and operationalised as required for the study. The criteria adopted for considering the major components to be selected were:

1. The degree of relevancy of the sub components to measure sustainable livelihood through agro-biodiversity conservation.
2. The suitability of the sub components to the study area.
3. The extent of quantifiability and measurability of the sub components.
4. The data availability on the sub components.
5. The clarity and specificity of the sub components.
6. Avoiding the duplication of sub components.

Table 3.4. Components Identified and Selected after Factor Analysis.

Sl.No.	*Major Component*	*No. of Sub Components*
1.	Food sustainability	7
2.	Nutritional security	6
3.	Health management	5
4.	Shelter security	5
5.	Soil and water sustainability	5
6.	Economic sustainability	5
7.	Employment generation	4
8.	Educational security	4
9.	Assets management	3
10.	Cultural sustainability	4
11.	Infrastructure and marketing sustainability	4
12.	Political sustainability	4
	Total	**56**

Selection of Sub Components

The sub components for each of the major component were identified. There were 80 sub components from which 56 components were selected by sending to the same set of judges. As the number of sub components was high it was decided to select only such of those very important sub components by applying factor analysis. The statements were selected for the study if the factor value is above 0.70 (Appendix VI).

Development of Sustainable Livelihood Index (SLI)

Definition for Index

The term index has been operationalised as a numerical scale used to compare variables with one another. It is a statistical indicator providing a representation of the value of the securities which constitute it. Indices often serve as a barometer for a given market or industry and benchmarks against which financial or economic performance is measured.

Procedure for Index Development

Sustainable Livelihood Index (SLI) was a systematic approach to link the issues of improvement of standard of living, sustainability and development process. It was arrived by adding the total scores of each component *viz.*, Food sustainability, nutritional security, health management, shelter security, soil and water sustainability, economic sustainability, employment generation, educational security, assets management, cultural sustainability, infrastructure and marketing sustainability and political sustainability. It has been calculated by adopting the following mathematical formula.

$$\text{Sustainable Livelihood Index (SLI)} = \frac{(I_{1X} + I_{2X} + I_{3X} + \dots\dots\dots I_{12X})}{(I_{1Y} + I_{2Y} + I_{3Y} \dots\dots\dots I_{12Y})} \times 100$$

where,

I_{1X}: Total score obtained by the 1st component

I_{2X}: Total score obtained by the 2nd component

I_{12X}: Total score obtained by the 12th component

I_{1Y}: Total maximum possible score for 1st component by an individual

I_{2Y}: Total maximum possible score for 2nd component by an individual

I_{12}: Total maximum possible score for 12th component by an individual

Food Sustainability Index (FSI)

Food is the basic of all human needs. When a biodiversity system could supply the food need of the community it was comparatively better in nature. The timely availability of food, adequacy in supply, the easy availability of food were some of the important sub components of food sustainability.

FSI was worked out by using the formula

$$FSI = \frac{A\,FS_{Xi}}{T\,FS_{Yi}} \times 100$$

FSI: Food Sustainability Index

$A\,FS_{Xi}$: Actual score obtained by an individual on food sustainability

$T\,FS_{Yi}$: Total possible maximum score for an individual on food sustainability

Nutritional Security Index (NSI)

Food alone was not sufficient in the recent days, whereas nutritional food was always essential. This referred to the nutritional supply of the agro-biodiversity to the community. It could be verified by the supply of basic nutrition, essential mineral and vitamins supply, uniform supply of nutrition, easy availability of nutrition and support of livestock for nutritional supply.

NSI was worked out by using the formula

$$NSI = \frac{A\,NS_{Xi}}{T\,NS_{Yi}} \times 100$$

NSI: Nutritional Security Index

$A\,NS_{Xi}$: Actual score obtained by an individual on nutritional security

$T\,NS_{Yi}$: Total possible maximum score for an individual on nutritional security

Health Management Index (HMI)

Livelihood sustainability was always linked with health management practices. Hence HMI referred to the health management components. It was argued with the supply of healthy food and medicinal valued crops in traditional and cultivated varieties. It showed the essentiality of medicinal plants in health management practices.

HMI was worked out by using the formula

$$HMI = \frac{A\,HM_{Xi}}{T\,HM_{Yi}} \times 100$$

HMI: Health Management Index

$A\,HM_{Xi}$: Actual score obtained by an individual on health management

$T\,HM_{Yi}$: Total possible maximum score for an individual on health management

Shelter Security Index (SSI)

Shelter was another important component of sustainable livelihood. Hence shelter security referred that the way by which the agro-biodiversity could satisfy the shelter needs of the community. It was measured by the materials available to build house, easy availability, traditional building methods and adequacy of supply of the building materials.

SSI was worked out by using the following

$$SSI = \frac{A\,SS_{Xi}}{T\,SS_{Yi}} \times 100$$

SSI: Shelter Security Index

$A\,SS_{Xi}$: Actual score obtained by an individual on shelter security

$T\,SS_{Yi}$: Total possible maximum score for an individual on shelter security

Soil and Water Sustainability Index (SWI)

Soil and water were the essential components of sustainable livelihood. Hence soil and water sustainability index referred to how the agro-biodiversity conservation practices could maintain soil and water for sustainable livelihood. It was measured by the quality of soil and water, traditional practices essentiality and health of soil and water.

SWSI was worked out by using the following formula

$$SWSI = \frac{A\,SWS_{Xi}}{T\,SWS_{Yi}} \times 100$$

SWSI: Soil and Water Sustainability Index

$A\,SWS_{Xi}$: Actual score obtained by an individual on soil and water sustainability

$T\,SWS_{Yi}$: Total possible maximum score for an individual on soil and water sustainability

Economic Sustainability Index (ESI)

Economic sustainability index referred to the economic supply and viability of the agro-biodiversity system to the community. It was the economic sustenance, viability, monetary benefits, economic value of the traditional crops and forest produces.

ESI was worked out by using the following formula

$$ESI = \frac{A\,ES_{Xi}}{T\,ES_{Yi}} \times 100$$

ESI: Economic Sustainability Index

$A\,ES_{Xi}$: Actual score obtained by an individual on economic sustainability

$T\,ES_{Yi}$: Total possible maximum score for an individual on economic sustainability

Employment Generation Index (EGI)

Employment generation referred to the employment opportunities given by the agro-biodiversity to the community. It was measured by the adequacy of supply, timeliness, sufficiency to the members and mental satisfaction.

EGI was worked out by using the following

$$EGI = \frac{A\,EG_{Xi}}{T\,EG_{Yi}} \times 100$$

EGI: Employment Generation Index

$A\,EG_{Xi}$: Actual score obtained by an individual on employment generation

$T\,EG_{Yi}$: Total possible maximum score for an individual on employment generation

Educational Security Index (ESI)

Educational security referred to the educational background a community could obtain from the agro-biodiversity measures. Quality and lifelong education was the important component in sustainable livelihood.

ESI was worked out by using the following formula

$$ESI = \frac{A\,ES_{Xi}}{T\,ES_{Yi}} \times 100$$

ESI: Educational Security Index

$A\,ES_{Xi}$: Actual score obtained by an individual on educational security

$T\,ES_{Yi}$: Total possible maximum score for an individual on educational security

Assets Management Index (AMI)

Assets management referred to the way through which livelihood sustainability can be attained by the development of human capital, socio capital and economic capital.

AMI was worked out by using the following formula

$$AMI = \frac{A\,AM_{Xi}}{T\,AM_{Yi}} \times 100$$

AMI: Assets Management Index

$A\,AM_{Xi}$: Actual score obtained by an individual on assets management

$T\,AM_{Yi}$: Total possible maximum score for an individual on assets management

Cultural Sustainability Index (CSI)

Cultural sustainability referred to the maintenance of cultural heritage, enriching it and maintaining its importance towards sustainable livelihood.

CSI was worked out by using the following formula

$$CSI = \frac{A\,CS_{Xi}}{T\,CS_{Yi}} \times 100$$

CSI: Cultural Sustainability Index

$A\,CS_{Xi}$: Actual score obtained by an individual on cultural sustainability

$T\,CS_{Yi}$: Total possible maximum score for an individual on cultural sustainability

Infrastructure and Marketing Sustainability Index (IMI)

Infrastructure and marketing index referred to the importance of infrastructure and good market price for the produce, which in turn made a better livelihood.

IMI was worked out by using the following formula

$$IMI = \frac{A\,IM_{Xi}}{T\,IM_{Yi}} \times 100$$

IMI: Infrastructure and Marketing Index

$A\,IM_{Xi}$: Actual score obtained by an individual on infrastructure and marketing

$T\,IM_{Yi}$: Total possible maximum score for an individual on infrastructure and marketing

Political Sustainability Index (PSI)

Political sustainability referred to the government policies and political system that encouraged the traditional practices and collection of forest produces.

PSI was worked out by using the following formula

$$PSI = \frac{A\,PS_{Xi}}{T\,PS_{Yi}} \times 100$$

PSI: Political Sustainability Index

$A\,PS_{Xi}$: Actual score obtained by an individual on political sustainability

$T\,PS_{Yi}$: Total possible maximum score for an individual on political sustainability

3.5. Techniques of Data Collection

Interview Schedule

Taking into consideration the scope and objectives of the study a comprehensive interview schedule was prepared. The schedule consisted of three parts. It consisted of the profile characteristics, questions related to participation pattern of tribal women in agro-biodiversity conservation and sustainable livelihood through agro-biodiversity conservation. The pre-testing of the interview schedule was done with 30 respondents selected from non-sampling area. Refinement was made in the schedule after pretesting as found necessary and it was finalised (Annexure VII). Precaution was taken to ensure that the items were clear, concise, complete, comprehensive, and unambiguous. Data regarding the profile of the respondents, questions related perception, participation pattern and sustainable livelihood were collected through the interview schedule. Figure 3.2 depicts the conceptual model explaining the major issues of the study.

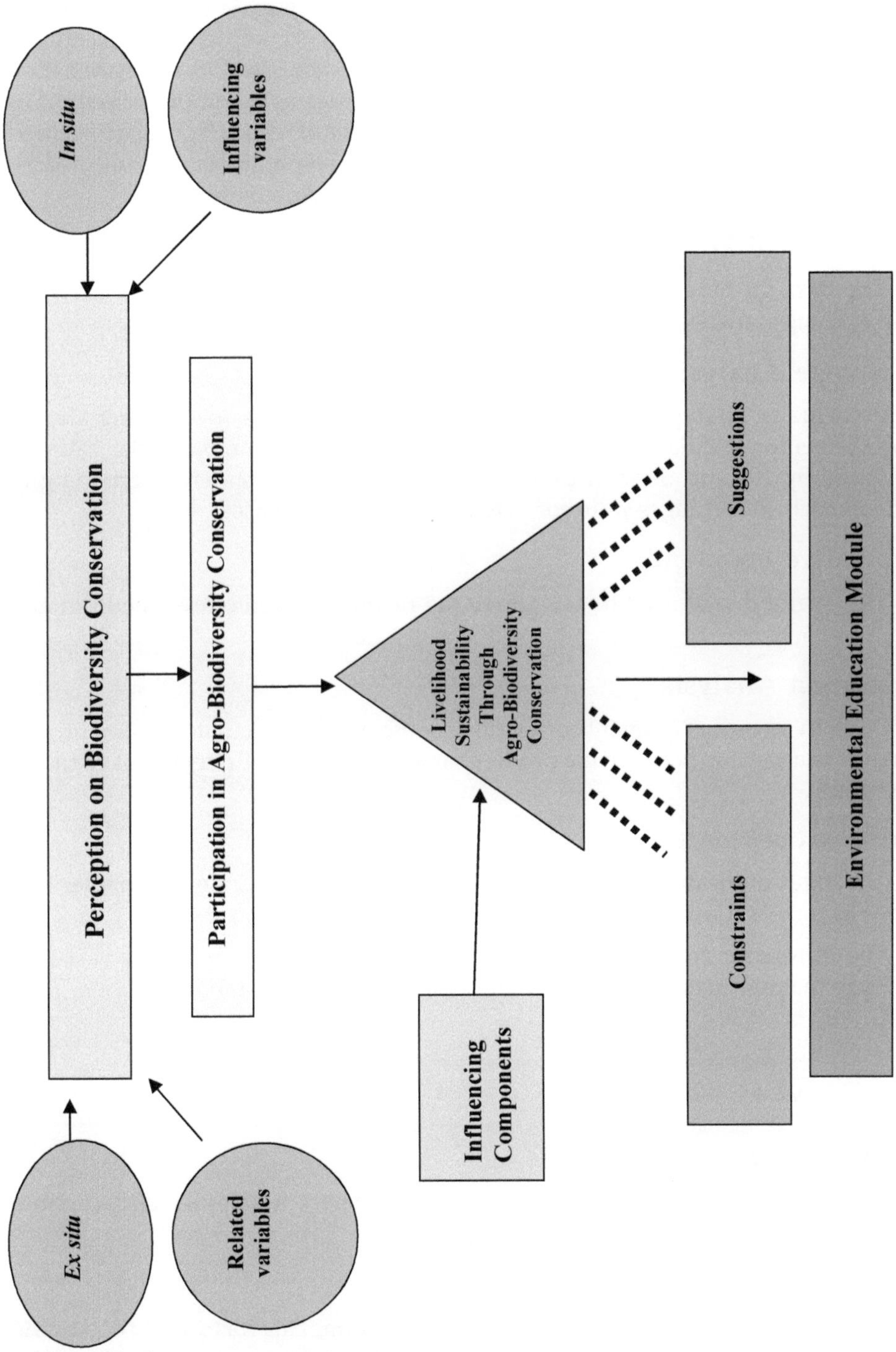

Figure 3.2. Conceptual Model of the Study.

Case Analysis

Case analysis was an empirical inquiry that investigated a contemporary phenomenon within its real life context. It was a method of comprehensive study of a social unit, which may be a person, a family, a group, an institution, an organization or a community. The case analysis may itself be used as a complete method of investigation or may be used in early exploratory stages of research or supplemented in other research evidences to add depth to them. Two respondents were selected for conducting case analysis. The case studies are given in Annexure VIII.

3.6. Statistical Tools Used

The data gathered were quantified and tabulated for statistical analysis. The following statistical techniques were applied for the study.

Percentage Analysis

Percentage analysis was used in descriptive analysis for making simple comparisons for calculating percentages. The frequency of the particular cell was multiplied by 100 and divided by the total number of respondents pertaining to particular cell. Percentages were corrected to two decimal places.

Cumulative Frequency

This method was used to categorize the respondents into low, medium and high.

Correlation Analysis

Pearson's product moment correlation coefficient was calculated to find out the degree of relationship between the profile characteristics and the dependent variables of the study.

Correlation Coefficient

The index of the degree of relationship between two continuous variables was known as correlation coefficient.

The correlation coefficient, r is given as the ratio of covariance of the variables x and y to the product of the standard deviations of x and y symbolically.

$$r = \frac{\Sigma xy - \Sigma x \ \Sigma y/n}{\sqrt{\Sigma x^2 - (\Sigma x)^2/n}\sqrt{\Sigma y^2 - (\Sigma y)^2/n}}$$

Step-wise Regression Analysis

Step-wise regression analysis was used to know the variables which contribute much towards the dependent variable.

Factor Analysis

Factor analysis was done using the varimax rotation. This was done to determine the number and nature of factors underlying the dependent variable. Based on the factor loadings, the sub components were selected under each major component.

Chapter 4
Findings and Discussion

This chapter highlights the findings of the investigation in accordance with the objectives set forth along with suitable discussion under the following sections:

4.1. Profile characteristics of tribal women

4.2. Perception on agro-biodiversity conservation

4.3. Participation in agro-biodiversity conservation

4.4. Livelihood sustainability of the respondents through agro-biodiversity conservation

4.5. Constraints faced by the tribal women in participating in the agro-biodiversity conservation practices

4.6. Suggestions for sustainable livelihood through agro-biodiversity conservation

4.7. Environmental Education Module (EEM) for sustainable livelihood through agro-biodiversity conservation

4.1. Profile Characteristics of Tribal Women

Age

The chronological completed years of age was collected from the respondents and it was classified as young (less than 30 years), middle (31-45years) and old (more than 45 years). The results are presented in the Table 4.1.

Table 4.1. Distribution of Tribal Women According to Age.

Sl.No.	Category	Kothar (n=100)		Irular (n=100)		Total (n=200)	
		No	Per cent	No	Per cent	No	Per cent
1.	Young (less than 30 years)	18	18.00	24	24.00	42	21.00
2.	Middle (31-45 years)	52	52.00	58	58.00	110	55.00
3.	Old (more than 45 years)	30	30.00	18	18.00	48	24.00

From the results depicted in Table 4.1 it was noted that nearly half (52.00 per cent) of the Kothars were of middle aged group followed by old (30.00 per cent) and young (18.00 per cent) aged group. In the same way majority of the Irulars were middle aged group (58.00 per cent) followed by young (24.00 per cent) and old (18.00 per cent) aged group. Among the total respondents 55.00 per cent of the respondents were middle aged group followed by old (24.00 per cent) and young (21.00 per cent) aged group. The respondents of the young aged group were actively involved in other income generating occupations, such as employment in private companies and construction work. Some of the families had been migrated from the village to nearby town in search of income. The old aged respondents were also not involved much in the field activities. Hence the active groups involved in the agro-biodiversity conservation practices were the middle aged group.

The middle aged groups were the farm and home makers of each household. It was the right group where in the new technologies could be taught and a great turnover could be made to create some impact on the community. They were also friendly with each other and share the important issues among them. This was useful for the respondents to actively participate in agro-biodiversity conservation practices. The results are in line with the findings of Marimuthu (2001) and Bhuvaneswari (2005) wherein they reported that majority of the respondents were middle aged group.

Educational Status

According to the results in Table 4.6 and Figure 4.1 it was learnt that 40.00 per cent of the Kothars were educated upto secondary school and then equally shared by primary (23.00 per cent) and middle (23.00 per cent) school education followed by the illiterate group 10.00 per cent. Only four per cent of the respondents were of higher secondary education. None were in the collegiate group. In the case of the Irulars 42.00 per cent of the respondents were of primary school education followed by secondary (30.00 per cent) illiterate (15.00 per cent), middle school education (11.00 per cent) and Higher Secondary (2.00 per cent). Among the overall classification of the respondents 35.00 per cent were of secondary school education followed by primary (32.50 per cent), middle (17.00 per cent), illiterate (12.50 per cent) and higher secondary (3.00 per cent). None of the respondents' possessed collegiate education. The Kothar habitation was located nearby the town; hence they could acquire secondary school education from government higher secondary school and also from the private schools and illiterates were very meager in number. They could

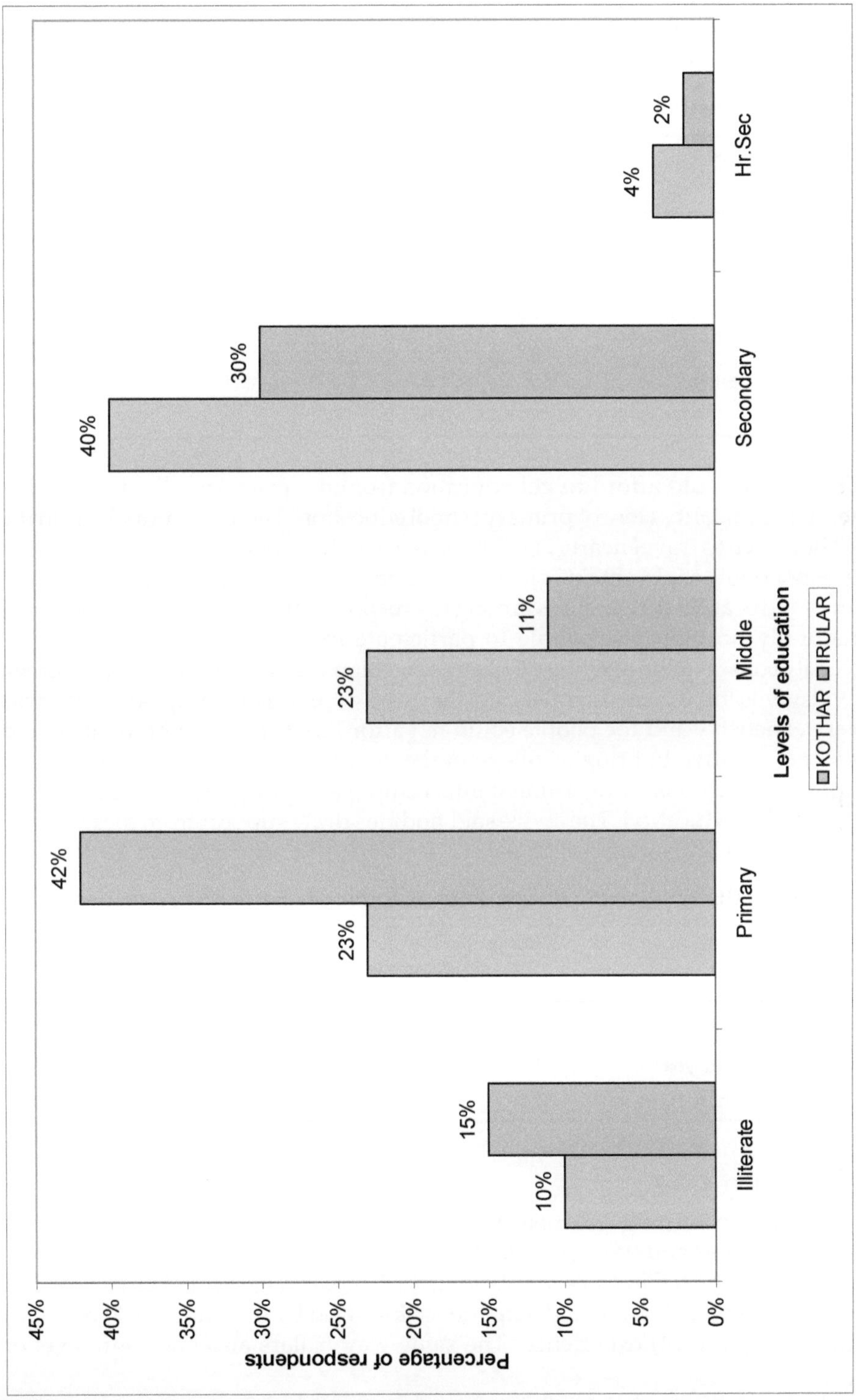

Figure 4.1. Distribution of the Respondents according to their Educational Status.

afford upto middle school education from their own village. But the Irular habitation was located far away from the town. Naturally it prevented the women to travel long for higher education.

Table 4.2. Distribution of Tribal Women according to Educational Status.

Sl.No.	Category	Kothar (n=100)		Irular (n=100)		Total (n=200)	
		No	Per cent	No	Per cent	No	Per cent
1.	Illiterate	10	10.00	15	15.00	25	12.50
2.	Primary	23	23.00	42	42.00	65	32.50
3.	Middle	23	23.00	11	11.00	34	17.00
4.	Secondary	40	40.00	30	30.00	70	35.00
5.	Higher Secondary	4	4.00	2	2.00	6	3.00
6.	Collegiate	0	0.00	0	0.00	0	0.00

The Irulars could afford to get education from the primary school from their village. Hence majority were of primary school education. There was no college in the town. They have to travel nearly 20-25 kms to reach the colleges of the nearby town. So the respondents could obtain college education. The needed steps should be taken to promote educational awareness among the respondents. More educational status will naturally promote the people to participate in the environment and agro-biodiversity conservation activities. College education was still a dream to the women in the village, as there were no colleges in the nearby areas. Traveling a long way was also very expensive and the people could not afford to it. The Government should make some initiative to bring a college in the nearby town and encourage higher education to the women. Government intervention and policy implications should reflect on higher education. The above said findings draw support from the evidence of the findings of Bhuvaneswari (2005).

Table 4.3. Distribution of Tribal Women according to Farming Experience.

Sl.No.	Category	Kothar (n=100)		Irular (n=100)		Total (n=200)	
		No	Per cent	No	Per cent	No	Per cent
1.	Low (less than 10 years)	7	7.00	18	18.00	25	12.50
2.	Medium (10-15 years)	23	23.00	12	12.00	35	17.50
3.	High (more than 15 years)	70	70.00	70	70.00	140	70.00

Farming Experience

Experience in farming was measured in total years the individual had indulged in farming activities and the respondents were classified as low, medium and high. The results are depicted in Table 4.3. Among the Kothars majority (70.00 per cent) of the respondents were having high experience followed by medium (23.00 per cent) and low (7.00 per cent) experience. The same way Irulars also had high level of experience (70.00 per cent) followed by low (18.00 per cent) and medium (12.00 per cent) levels. Considering the sum total nearly 70.00 per cent of the respondents had high level of farming experience, followed by medium (17.50 per cent) and low (12.50 per cent) levels of experience.

The experience in farming could be said as the experience where-in the respondents concentrate much on the agro-biodiversity conservation issues. The tribal women were the active participants in farming. They were more interested in organic farming as well as the traditional farming. They supported the cultivation of the minor millets and the diversified cropping pattern. Community level experience in farming would work well in due course of time. The higher level of experience revealed that the main occupation of the tribal women was farming. High level of experience in farming would create a positive attitude among the respondents and make them actively participate in the agro-biodiversity conservation activities. The higher level of farming experience of the tribal women derives support from the research findings of Marimuthu (2001).

Social Participation

Social participation of the respondents was measured as the membership of the respondents in various organizations such as cooperative societies, cooperative banks, mahila mandals etc. The respondents were classified as low, medium and high according to their level of participation. From Table 8 it was studied that more than half of the Kothars (52.00 per cent) were having medium level of social participation followed by high (30.00 per cent) and low (18.00 per cent) levels. In the same way majority of the Irular's (58.00 per cent) were having medium level of social participation followed by low (24.00 per cent) and high (18.00 per cent) levels of participation. Among the overall respondents 55.00 per cent of the respondents were having medium level of social participation followed by 24.00 per cent with high level and 21.00 per cent with low level of social participation.

Table 4.4. Distribution of Tribal Women according to Social Participation.

Sl.No.	*Category*	*Kothar (n=100)*		*Irular (n=100)*		*Total (n=200)*	
		No	*Per cent*	*No*	*Per cent*	*No*	*Per cent*
1.	Low	18	18.00	24	24.00	42	21.00
2.	Medium	52	52.00	58	58.00	110	55.00
3.	High	30	30.00	18	18.00	48	24.00

Social participation was an indicator where in the commitment of the individual towards the society upliftment was measured. The respondent joined as a member or an office bearer in various organizations, in order to widen their contacts and way of living.

It was observed from the Table 4.5 that invariably all the respondents were members in the self help groups in their habitations. It was also noticed that majority of (70.00 per cent) of the Kothars were members in the milk co-operatives and one-third (32.00 per cent) of them were members in the co-operative banks. But in the case of Irulars it was different. Nearly half (44.00 per cent) of the respondents were members in the milk co-operatives followed by 40.00 per cent of them being members in the co-operative bank. Excepting for a meager per cent of Kothars (22.00 per cent) and Irulars (8.00 per cent) were members in the Mahila Mandals. Self help group as an institution was fully developed in these villages. Effective functioning of these group activities enables the respondents for community participation in agro-biodiversity conservation practices. Figure 4.2 explains the distribution of the respondents according to membership in organizations.

Table 4.5. Distribution of Tribal Women according to Membership in Organizations.

Sl.No.	Organisation	Kothar (n=100)*		Irular (n=100)*		Total (n=200)*	
		No	Per cent	No	Per cent	No	Per cent
1.	Self Help Groups	100	100.00	100	100.00	200	100.00
2.	Milk Co-operatives	70	70.00	44	44.00	114	57.00
3.	Co-op banks	32	32.00	40	40.00	72	36.00
4.	Mahila Mandals	22	22.00	8	8.00	30	15.00

*: Multiple response.

These types of organizations build up the values of the society and the environment. Hence higher level of social participation would promote the respondents to participate in agro-biodiversity conservation activities. Most of the respondents participated in self help group activities as members and organizers. They promoted the conservation measures in the community level. Local communication also had increased and the women were participating in the district and the state level exhibitions. This type of sociological approach increased the community level participation in agro-biodiversity conservation. This result is in line with the results of Marimuthu (2001).

Information Seeking Behaviour

Human beings always rely on other human or some other sources for information sharing and receiving. The information seeking behaviour of the individual was measured as the information received from various sources such as friends, relatives, neighbours, and the frequency of information received from these sources regarding the agro-biodiversity conservation. The respondents were classified as low, medium and high levels. The results are depicted in Table 4.6.

Table 4.6. Distribution of Tribal Women according to Information Seeking Behaviour.

Sl.No.	Category	Kothar (n=100)		Irular (n=100)		Total (n=200)	
		No	Per cent	No	Per cent	No	Per cent
1.	Low	4	4.00	12	12.00	16	8.00
2.	Medium	65	65.00	83	83.00	148	74.00
3.	High	31	31.00	5	5.00	36	18.00

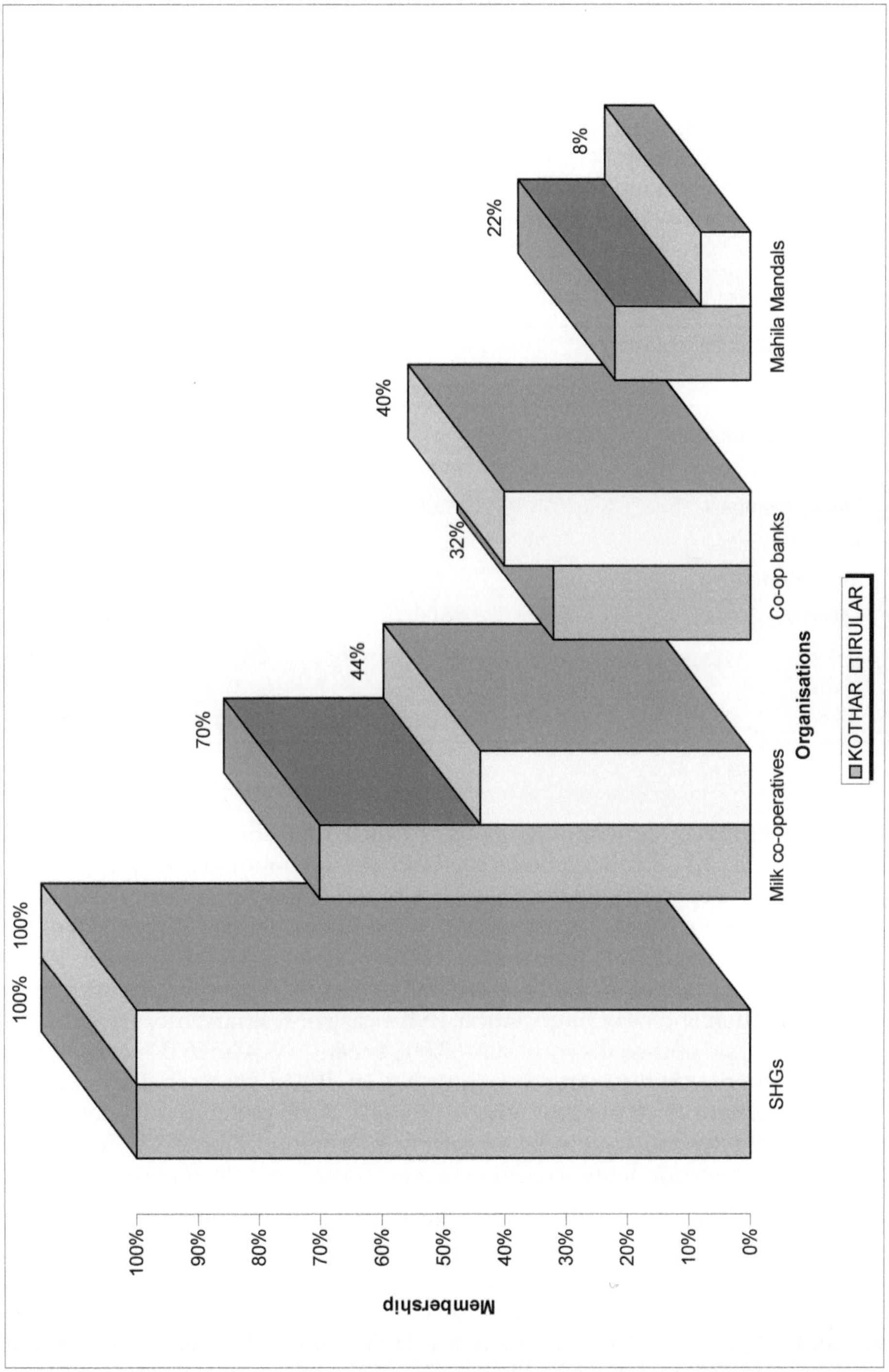

Figure 4.2. Distribution of the Respondents according to Membership in Organisations.

Majority of Kothars were having medium (65.00 per cent) level of information seeking behaviour followed by high (31.00 per cent) level. Only 4.00 per cent of the respondents were having low level of information seeking behaviour. With respect to Irulars majority of the respondents (83.00 per cent) were having medium level of information seeking behaviour followed by low level (12.00 per cent). Only 5.00 per cent of the respondents had high level of information seeking behaviour. Among the overall respondents, nearly 74.00 per cent of the respondents were having medium level of information seeking behaviour followed by 18.00 per cent having high and 8.00 per cent having low level of information seeking behaviour. The spread of the respondents were between medium to high level of information seeking behaviour.

Table 4.7. Distribution of tribal women according to sources of information

Sl.No.	Sources of Information	Kothar (n=100)*		Irular (n=100)*		Total (n=200)*	
		No	Per cent	No	Per cent	No	Per cent
	Non institutional						
1.	Neighbours	75	75.00	60	60.00	135	67.50
2.	Family members	71	71.00	59	59.00	130	65.00
3.	Relatives	70	70.00	52	52.00	122	61.00
	Institutional						
4.	Officials	63	63.00	45	45.00	108	54.00
5.	NGOs	60	60.00	40	40.00	100	50.00
	Media						
6.	Radio	60	60.00	32	32.00	92	46.00

* Multiple responses.

The distribution of the respondents according to their sources of information is depicted in Table 4.7. While considering both the respondents, majority of the respondents were depending on their neighbours and fellow farmers for their source of information. Three-fourth (75.00 per cent) of the Kothars and 60.00 per cent of the Irulars depend on neighbours followed by family members for information. It was found that 70.00 per cent of the Kothars and half of the (52.00 per cent) Irulars rely on the relatives as their choice of information. In the case of Kothars, more than half of the respondents had opined that officials (63.00 per cent), NGOs (60.00 per cent) and radio (60.00 per cent) as their source of information. It was somewhat lesser in the case of Irulars where 45.00 per cent rely on officials. Forty per cent consider NGOs and 32.00 per cent depended on radio for getting information. Figure 4.3 explains the distribution of the respondents according to their sources of information.

The respondents could share the information among their relatives and neighbours. Friends were not much influential. Being in a tribal community there were much restrictions to share their information with the other community or outside their locality. The extension scientists of UPASI and the officials of the State Department of Agriculture were good sources of information. Most of the respondents were having good contact with the local NGOs. The programmes organised by the

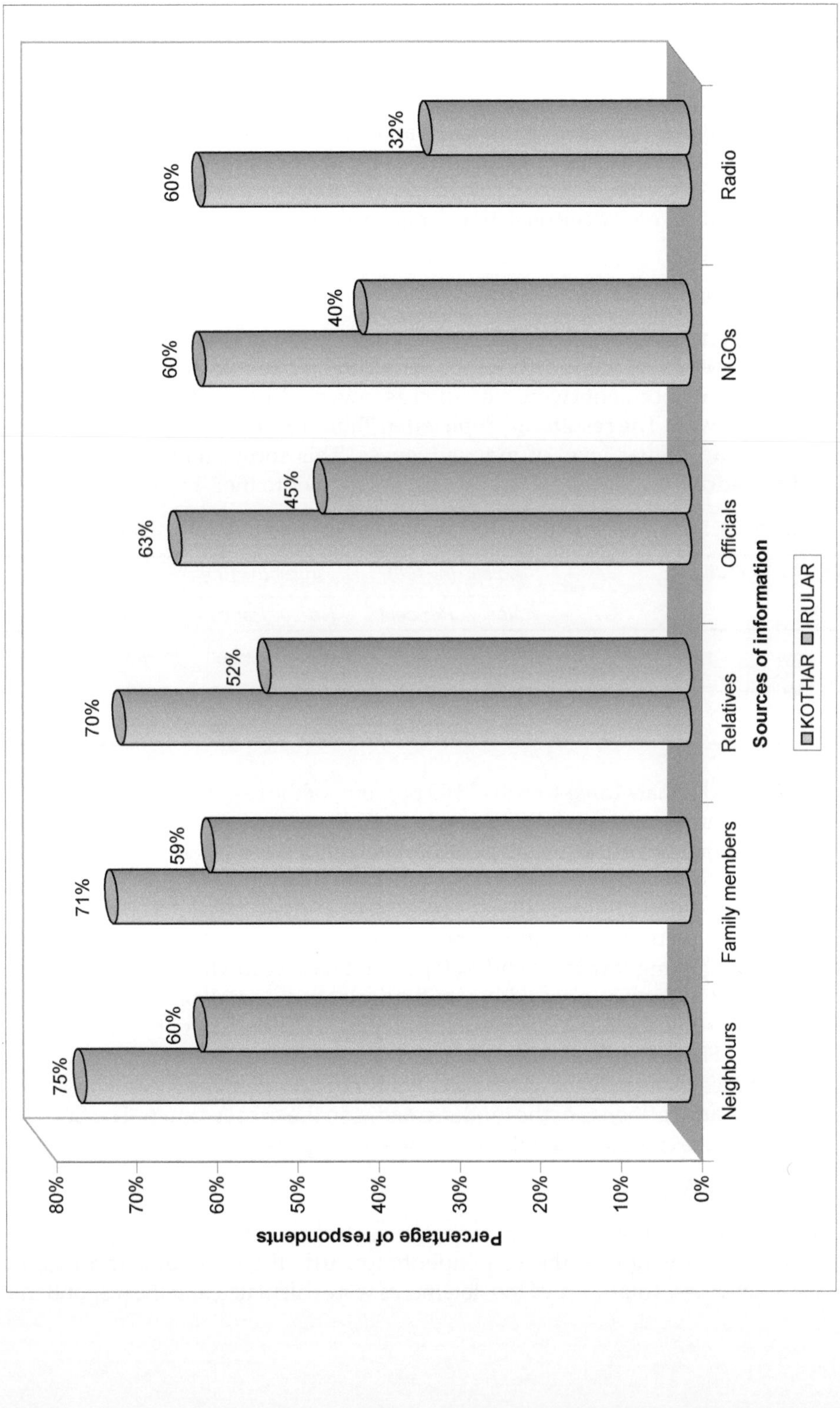

Figure 4.3. Distribution of the Respondents according to Sources of Information.

NGO such as drama, exhibitions, tours and seminars were also good sources of information. The mass media such as radio played a very vital role in the information giving arena.

These kind of programmes created awareness among the respondents regarding the agro-biodiversity loss and consequences. It promoted the feeling to safeguard the biodiversity and the valuable resources available. Therefore information seeking behaviour had high influence towards the participation of the respondents in agro-biodiversity conservation. In turn it also helped in attaining livelihood sustainability through agro-biodiversity conservation. This finding is in line with the findings of Bhuvaneswari (2005).

Innovativeness

Innovativeness was measured in terms of the statements that at what period an individual agrees to adopt a technology. The time lapse was measured in three statements. The respondents were classified as low, medium and high according to the responses given. The results are depicted in Table 4.8. Invariably majority of the respondents had higher level of innovativeness. This shows that majority of the respondents adopted the technology as soon as it comes to their knowledge.

Table 4.8. Distribution of the Tribal Women according to Innovativeness.

Sl.No.	Category	Kothar (n=100)		Irular (n=100)		Total (n=200)	
		No	Per cent	No	Per cent	No	Per cent
1.	Low	17	17.00	42	42.00	59	29.50
2.	Medium	8	8.00	2	2.00	10	5.00
3.	High	75	75.00	56	56.00	131	65.50

Among the Kothars three-fourth (75.00 per cent) of the respondents were having higher level of innovativeness followed by low (17.00 per cent) and medium (8.00 per cent) level. Majority of the Irular's (56.00 per cent) were having high level of innovativeness followed by low (42.00 per cent) and medium (2.00 per cent) level of innovativeness. Among the overall respondents, nearly two-third (65.00 per cent) of the respondents were having high level of innovativeness followed by one-third (30.00 per cent) having low level and 5.00 per cent having medium. Because of the higher level of information seeking behaviour and social participation, the respondents were aware about the new technologies and also have got the motivation to adopt them. Higher level of innovativeness always influence the participation all types of activities, especially the agro-biodiversity conservation practices. This shows the motivation and influence of the modern media on the tribal women. The findings of Marimuthu (2001) coincided with the results of the present study.

Economic Motivation

Economic motivation of the respondents was studied with the help of statements revealing the motivation of the respondents towards the economic and income generating activities. Total score of the statements were calculated and the respondents were classified as low, medium and high. The results are depicted in Table 4.9. Nearly half of the Kothars were having medium level of economic motivation (54.00 per cent) followed by high (39.00 per cent) and low (7.00 per cent) levels of economic motivation. Nearly half (49.00 per cent) of the Irulars had higher level of economic motivation followed by medium (37.00 per cent) and low (14.00 per cent) levels. Among the overall respondents equal amount of distribution among the medium (46.00 per cent) and high (44.00 per cent) level of economic motivation was found. Only 10.00 per cent of the respondents had low level of economic motivation. The spread of the respondents according to economic motivation was between medium to high. Economic motivation was the overall attitude of the respondents towards economic returns. Value addition activities and marketing facilities of the traditional crops increased the economic motivation of the tribal women.

Table 4.9. Distribution of Tribal Women according to Economic Motivation.

Sl.No.	*Category*	*Kothar (n=100)*		*Irular (n=100)*		*Total (n=200)*	
		No	*Per cent*	*No*	*Per cent*	*No*	*Per cent*
1.	Low	7	7.00	14	14.00	21	10.50
2.	Medium	54	54.00	37	37.00	91	45.50
3.	High	39	39.00	49	49.00	88	44.00

Hard work and the employment always results in monetary benefit of the respondents. Any activity, technology or variety would be easily accepted and adopted by the respondents when it provided some economic benefit to the individual. High level of economic motivation influenced the participation pattern of the individual in agro-biodiversity conservation. When conserved the biodiversity provides good economic return to the individual. It provides an opportunity to cultivate variety of crops and shifting cropping pattern. Traditional varieties were boon to the mankind. They fetched good amount of return in the market. Hence the participation of tribal women in agro-biodiversity conservation was very high when considering economic motivation. This result could be supported from the evidence of Marimuthu (2001) and Bhuvaneswari (2005).

Progressiveness

Progressiveness was measured by the advanced thinking of the individual. Their motivation on their future and how far their needs were attained could be measured by the statements measured to give motivation on progressiveness. The respondents were classified as low, medium and high according to progressiveness.

Table 4.10. Distribution of Tribal Women according to Progressiveness.

Sl.No.	*Category*	*Kothar (n=100)*		*Irular (n=100)*		*Total (n=200)*	
		No	*Per cent*	*No*	*Per cent*	*No*	*Per cent*
1.	Low	21	21.00	11	11.00	32	16.00
2.	Medium	57	57.00	59	59.00	116	58.00
3.	High	22	22.00	30	30.00	52	26.00

Majority of the respondents (57.00 per cent) among the Kothars were having medium level of progressiveness followed by high (22.00 per cent) and low (21.00 per cent) levels. Among the Irulars more than half of the respondents (59.00 per cent) were of medium level followed by high (30.00 per cent) and low (11.00 per cent) level of progressiveness. Among the overall respondents majority (58.00 per cent) of the respondents were having medium level of progressiveness followed by high (26.00 per cent) and low (16.00 per cent) levels of progressiveness. It was evident from the data that most of the respondents were having medium to high level of progressiveness.

When there was high progressiveness the respondents would have a high level of social participation and this would lead towards a high level of participation in agro-biodiversity conservation. This made the respondents to fullfill their needs through agro-biodiversity. The results of medium and high levels of progressiveness could be supported from the findings of Marimuthu (2001).

Cosmopoliteness

Cosmopoliteness was the character of an individual which measured the outward looking of the individual. The respondents were classified as low, medium and high according to their level of cosmopoliteness. The results are depicted in Table 15. Three-fourth of the respondents among Kothars (75.00 per cent) were having high level of cosmopoliteness followed by medium (19.00 per cent) and low level (6.00 per cent) of cosmopoliteness. Majority of the Irulars (76.00 per cent) were having high level of cosmopoliteness followed by low (15.00 per cent) level. Only 9.00 per cent of the respondents were having medium level of cosmopoliteness. Considering the overall respondents three-fourth (75.00 per cent) of the respondents were having high level of cosmopoliteness followed by medium (14.00 per cent) and high (11.00 per cent) level. It was evident that high cosmopoliteness was felt among the respondents.

Table 4.11. Distribution of Tribal Women according to Cosmopoliteness.

Sl.No.	Category	Kothar (n=100)		Irular (n=100)		Total (n=200)	
		No	Per cent	No	Per cent	No	Per cent
1.	Low	6	6.00	15	15.00	21	10.50
2.	Medium	19	19.00	9	9.00	28	14.00
3.	High	75	75.00	76	76.00	151	75.50

Although the value system was different among the tribals cosmopoliteness was high. There was higher socialization which could have contributed towards cosmopoliteness. The taboos that were not needed had been broken. There was an essential transition towards the change process. Cosmopoliteness spoke about the outwardness of the individual and higher degree of contact towards the outer community. This type of information sharing was very much essential to ensure the awareness and adoption of recent technologies. The new ideas and information gathered from the outside community was very much essential for validation of the conservation measures. Hence the cosmopoliteness was an influential variable on the participation of the respondents in Agro-biodiversity conservation activities. This result could be evidenced from the results of Jamatia (1999).

Attitude Towards Agro-biodiversity Conservation

Attitude towards agro-biodiversity conservation was measured with the help of the statements assigned. The respondents were classified as unfavourable, somewhat fourable and highly favourable according to their level of attitude towards agro-biodiversity conservation.

Table 4.12. Distribution of Tribal Women according to Attitude Towards Agro-biodiversity Conservation.

Sl.No.	Category	Kothar (n=100)		Irular (n=100)		Total (n=200)	
		No	Per cent	No	Per cent	No	Per cent
1.	Unfavourable	6	6.00	14	14.00	20	10.00
2.	Somewhat favourable	30	30.00	6	6.00	36	18.00
3.	Highly favourable	64	64.00	80	80.00	144	72.00

Majority of the Kothars (64.00 per cent) were having highly favourable attitude towards agro-biodiversity conservation, followed by somewhat favourable (30.00 per cent) level. Only 6.00 per cent of the respondents were having unfavourable attitude towards agro-biodiveristy conservation. Majority of the Irulars were having highly favourable (80.00 per cent) attitude followed by unfavourable (14.00 per cent) and somewhat favourable (6.00 per cent) attitude. Among the overall respondents the spread was between somewhat favourable and highly favourable attitude towards agro-biodiversity conservation. This result coincided with the results of Marimuthu (2001) and Arunachalam (2002) where they had opined about the attitude levels of the clients towards environmental awareness.

Decision Making Behaviour

Decision making on the farm and home activities had been measured with the help of the activities and with whom the respondents consult for decision making. Mostly they consult their family members, friends and neighbours for deciding the farm and home activities. The results are depicted in Table 4.13.

Table 4.13. Distribution of Tribal Women according to Decision Making Behaviour.

Sl.No.	Category	Kothar (n=100)		Irular (n=100)		Total (n=200)	
		No	Per cent	No	Per cent	No	Per cent
1.	Low	2	2.00	19	19.00	21	10.50
2.	Medium	69	69.00	7	7.00	76	38.00
3.	High	29	29.00	74	74.00	103	51.50

Nearly three-fourth (69.00 per cent) of the Kothars had medium level of decision making behaviour followed by high level (29.00 per cent) and low level of (2.00 per cent) decision making behaviour. In the case of Irulars, three-fourth (74.00 per cent) of the respondents had high level of decision making behaviour followed by low (19.00 per cent) and medium (7.00 per cent) levels. Half (52.00 per cent) of the overall respondents had high level of decision making behaviour followed by medium (38.00 per cent) and low (10.00 per cent) levels of decision making behaviour.

The responses collected were based on the decision making level of the individuals as self, consulting with spouse or the elders of the family, consulting all the family members, helping others in decision making and no participation. The respondents were asked about their decision in selecting the crops to be grown, selection of varieties and cultivation of traditional crops.

Decision making behaviour reflected the level of participation of the tribal women in all kinds of agro-biodiversity conservation practices. When the decisions were self made they had higher impact on the individual. When the decisions were made with the consultancy of the family members it showed the family participation. In most of the cases decisions were collective and spoke about the community participation of agro-biodiversity conservation practices. Medium level of decision making behaviour of the farm women in farm and home practices could be evidenced from the findings of Premavathi (2002) and Bhuvaneshwari (2005).

The overall profile characteristics represented that the Kothars were somewhat developed than the Irulars in educational status, innovativeness and decision making behaviour. The other characteristics were on par among both the groups. While considering the attitude towards agro-biodiversity conservation the Irulars were more favourable towards agro-biodiversity conservation than the Kothars.

4.2. Perception on Agro-biodiversity Conservation

Perception was considered as a prerequisite to get an overall understanding of what the tribal women feel about agro-biodiversity conservation and thereby increase the participation in agro-biodiversity conservation. Hence ten dimensions were identified upon which the respondent's opinion was gathered.

Perception on *in situ* Conservation

Maintenance of Seeds of Traditional Crops

Perception of the respondents towards agro-biodiversity conservation was measured with two different components such as *in situ* and *ex situ* conservation. The statements were finalized and the responses were collected in three point continuum. The results are presented in the Table 4.14.

Table 4.14. Maintenance of Seeds of Traditional Crops.

Sl.No.	*Maintenance of Seeds*	*Kothar (n=100)*		*Irular (n=100)*	
		No	*Per cent*	*No*	*Per cent*
1.	Easy	45	45.00	39	39.00
2.	Somewhat easy	32	32.00	35	35.00
3.	Difficult	23	23.00	26	26.00

Maintenance of seeds of traditional crops in the field condition was an important *in situ* conservation measure. Less than fifty per cent of the Kothars revealed that it was easy to maintain seeds of traditional crops, whereas 32.00 per cent of them opined that it was somewhat easy and 23.00 per cent of them said that it was difficult to maintain the seeds of traditional crops. When there was crop rotation in the field the traditional seeds had to be secured for further use and rotation.

This was actually observed in almost all the villages visited during the data collection. Past studies by Marimuthu (2001) revealed that the clients could see the advantages of the traditional crops. The findings of this study lend support to the aforesaid findings of the earlier studies.

Traditional Crop Cultivation

The opinions expressed by the study groups on the traditional crop cultivation were obtained and presented in Table 4.15.

Table 19. Traditional Crop Cultivation.

Sl.No.	Traditional Crop Cultivation	Kothar (n=100)		Irular (n=100)	
		No	Per cent	No	Per cent
1.	Easy	54	54.00	50	50.00
2.	Somewhat easy	25	25.00	25	25.00
3.	Difficult	21	21.00	25	25.00

Apart from seed security the cultivation of the traditional crops also played a vital role in the *in situ* conservation. Nearly half of both the Kothars (54.00 per cent) and the Irulars (50.00 per cent) revealed that the cultivation of traditional crops were easy, only one- fourth of the respondents said that it was somewhat easy and difficult. When the respondents were interested in the security of traditional seeds it ultimately helped the respondents for cultivation of traditional crops. The field experiences also revealed that the tribal women were more interested in the cultivation of traditional crops rather that the scientific varieties. If proper guidance was provided along with marketing facility it would be a boon for their livelihood. This result was inline with the discussion of Bhuvaneshwari (2005).

Livelihood Sustainability through *in situ* Conservation

In situ conservation provided the livelihood sustainability through various avenues of employment and income generating avocations. How far it supported them was analysed. Two-third (60.00 per cent) of the Kothars perceived that *in situ* conservation fully supported their livelihood. In the same way 55.00 per cent of the Irulars also mentioned the same. About one-fourth (27.00 per cent) of the Kothars and 22.00 per cent of the Irulars had mentioned that the *in situ* conservation somewhat supported their livelihood. Only 13.00 per cent of the Kothars and 18.00 per cent of the Irulars felt that *in situ* conservation does not support their livelihood. The tribal groups led a life in harmony with the nature. Their basic needs were fully met by the nature. Figure 6 depicts the perception of the respondents on the support of livelihood sustainability on *in situ* conservation.

Table 4.16. Livelihood Sustainability through *in situ* Conservation.

Sl.No.	*Livelihood Sustainability*	*Kothar (n=100)*		*Irular (n=100)*	
		No	*Per cent*	*No*	*Per cent*
1.	Fully supported	60	60.00	55	55.00
2.	Somewhat supported	27	27.00	22	22.00
3.	Least support	13	13.00	18	18.00

The clients system which actually faced the hardships and constraints while deriving the benefits from the agro-biodiversity conservation through the set procedures expressed the existence of support for livelihood sustainability. Similar findings on the existence of complicated procedures were also reported by Marimuthu (2001).

Improvement in Standard of Living

As a result of the agro-biodiversity conservation over a period of time, the tribal women would have accrued benefits. Hence it was tested whether the respondents could acquired the benefits due to their participation in agro-biodiversity conservation practices. The data are presented in Table 4.17.

Table 4.17. Improvement in Standard of Living.

Sl.No.	*Improvement in Standard of Living*	*Kothar (n=100)*		*Irular (n=100)*	
		No	*Per cent*	*No*	*Per cent*
1.	Best improved	73	73.00	65	65.00
2.	Somewhat improved	27	27.00	27	27.00
3.	Least improved	10	10.00	18	18.00

To what extent the standard of living had been improved and how it had been perceived by the respondents was the theme of this analysis. Majority (73.00 per cent) of the Kothars perceived that there was best improvement in their standard of living. Nearly one-fourth of them perceived that there was some what improvement. Very meager per cent (10.00 per cent and 18.00 per cent) opined that it was least improvement. In the same way it was among the Irulars.

But on an average all the respondents accepted that the farmers could acquired the benefits due to agro-biodiversity conservation. It could be inferred that the respondents could not reap the fullest benefits out of the agro-biodiversity conservation practices. Formal steps could be taken to increase the benefits for the respondents. Marimuthu (2001) found the existence of benefits as perceived by the participants. So the findings obtained were corroborated with the findings of the earlier studies.

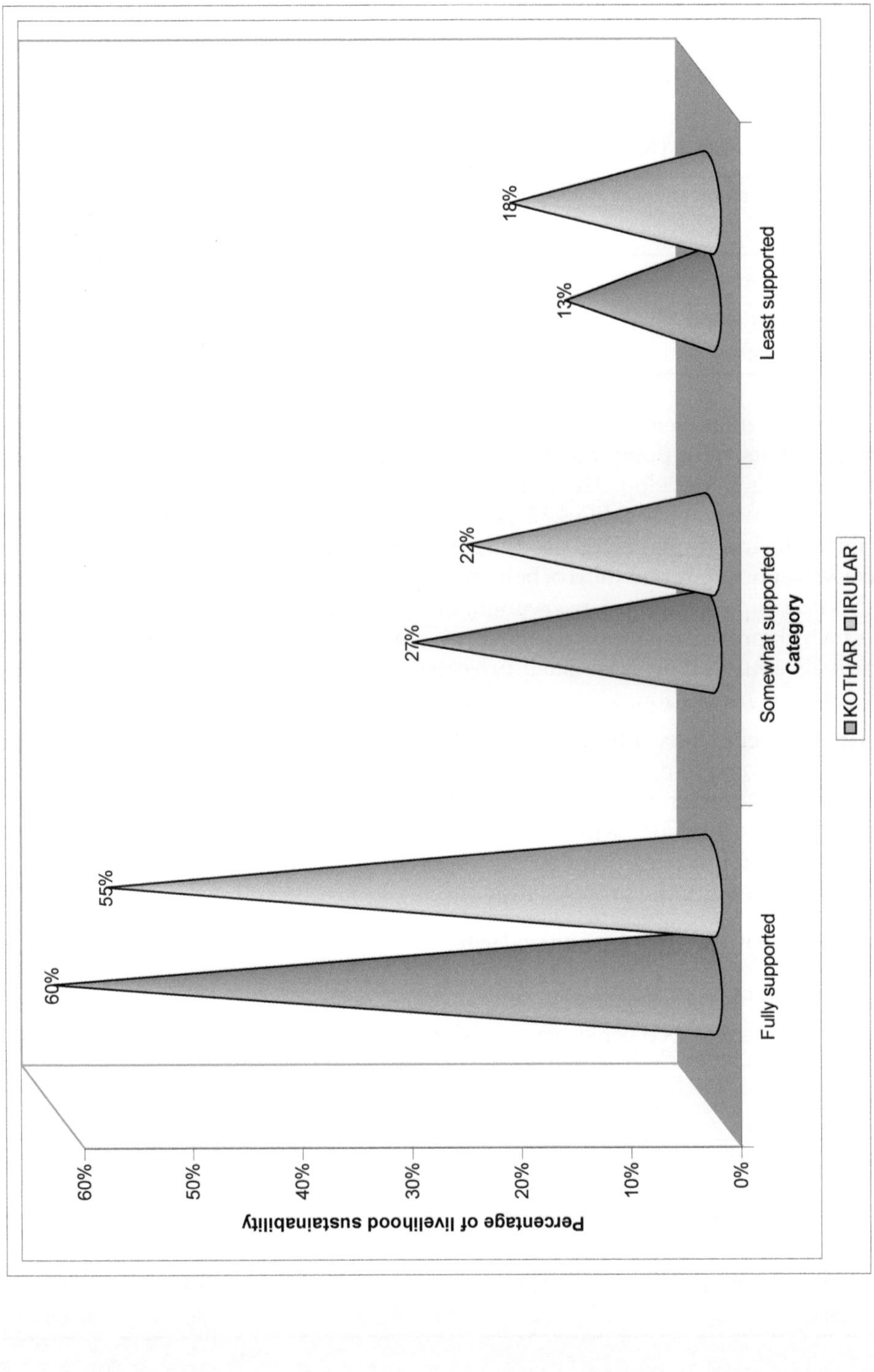

Figure 4.4. Perception on Livelihood Sustainability through *in situ* Conservation.

Government Policy for *in situ* Conservation

Keeping abreast of the latest know-how in the agro-biodiversity conservation was a prerequisite to cope up with the requirements of present day scientific agriculture. To know, how best the tribal women replenished their knowledge due to their participation in agro-biodiversity conservation practices the Government policies were analysed.

Table 4.18. Government Policy for *in situ* Conservation.

Sl.No.	Government Policy	Kothar (n=100)		Irular (n=100)	
		No	Per cent	No	Per cent
1.	Best suited	35	35.00	33	33.00
2.	Somewhat suited	45	45.00	42	42.00
3.	Least suited	20	20.00	25	25.00

Government policies were important for any development and participation of the respondents. If the policy measures were taken so keenly it would ultimately help the individuals to develop. The perception of the respondents on the government policies was given in the Table 4.18. Nearly 45.00 per cent of the Kothars and 42.00 per cent of the Irulars perceived that the government policies somewhat supported *in situ* conservation. Only one-third of both the group had mentioned that the government policies best supported the conservation. The joint forest management was one best project which provided support for the conservation of agro-biodiversity. The benefits that tribals could attain from such projects would motivate them towards agro-biodiversity conservation.

Government intervention for supporting the conservation practices provided a way for the people to participate in the conservation practices in a livelier manner. This gave an understanding of the agro-biodiversity conservation practices of varied nature to the respondents, which was largely needed to evoke their participation in the conservation practices. The findings reported by Bhuvaneshwari (2005) was also on the similar lines of the present study.

Perception on *ex situ* Conservation

Development of Seed Banks

Development of seed banks was one of the important *ex situ* conservation measure of the field.

Table 4.19. Development of Seed Banks.

Sl.No.	Development of Seed Banks	Kothar (n=100)		Irular (n=100)	
		No	Per cent	No	Per cent
1.	Easy	40	40.00	37	37.00
2.	Somewhat easy	20	20.00	23	23.00
3.	Difficult	40	40.00	40	40.00

Conserving the wild agro-biodiversity through seed banks was one tool with user-friendly nature and economic viability. The results are given in Table 23. By exchanging the seeds with the other communities the diversified crop cultivation was also ensured. Nearly 40.00 per cent of the Kothars and Irulars had perceived that the development of seed banks was very difficult. But the same level of respondents had mentioned that it was easy to develop seed banks. The eco-clubs in some of the community groups developed by the NGOs deliver the importance of seed banks. The genetic diversity of this particular area was measured with the help of these types of seed conservation efforts.

This was actually observed in almost all the villages visited during the data collection for the present study. As a result of closer interaction certainly it can be stated that nothing else would have contributed for their well-being other than agriculture, since they depended completely on farming. Past studies by Marimuthu (1998) could see the advantages of the traditional crops. The findings of this study lend support to the aforesaid finding of the earlier studies.

Traditional Food Habits Supported *ex situ* Conservation

The opinions expressed by the study groups on the traditional food habits were obtained and presented in Table 4.20.

Table 4.20. Traditional Food Habits Supported *ex situ* Conservation.

Sl.No.	*Traditional Food Habits*	*Kothar (n=100)*		*Irular (n=100)*	
		No	*Per cent*	*No*	*Per cent*
1.	Best supported	55	55.00	52	52.00
2.	Somewhat supported	31	31.00	30	30.00
3.	Least supported	14	14.00	18	18.00

In the recent years due to modernization the food habits of people had changed to a great extent. The nutritious minor millets and traditional crops could find no place in our diet. How far the traditional food habits supported *ex situ* conservation was measured. Nearly half (55.00 per cent) of the Kothars and Irulars (52.00 per cent) perceived that traditional food habits best supported *ex situ* conservation. Nearly one-third had mentioned that it somewhat supported. Only 14.00 per cent of Kothars and 18.00 per cent of the Irulars perceived that it least supported. Traditional food habits not only support food security, it also balanced the nutritional intake of the people. Ultimately it ensured a healthy diet apart from conserving the biodiversity. When the respondents were interested in the security of traditional seeds it ultimately helped the respondents for cultivation of traditional crops and led to traditional way of food habits. The field experiences also revealed that the tribal women were more interested in the cultivation of traditional crops rather than the scientific varieties. If proper guidance was provided for them with marketing facility it would be a boon for their upliftment. This result coincided with the discussion of Bhuvaneshwari (2005).

Livelihood Sustainability through *ex situ* Conservation

Ex situ conservation provided the livelihood sustainability through various avenues of employment and income generating avocations. How far it supported them was analysed. Two-third (65.00 per cent) of the Kothars perceived that *ex situ* conservation fully supported their livelihood. In the same way 58.00 per cent of the Irulars also mentioned the same. Nearly one-fourth (25.00 per cent) of the Kothars and 26.00 per cent of the Irulars had mentioned that the *ex situ* conservation somewhat supported their livelihood. Only 10.00 per cent of the Kothars and 16.00 per cent of the Irulars felt that *ex situ* conservation does not support their livelihood. The tribal groups lead a life in harmony with the nature. Their basic needs were fully met by the nature.

Table 4.21. Livelihood Sustainability through *ex situ* Conservation.

Sl.No.	Livelihood Sustainability	Kothar (n=100)		Irular (n=100)	
		No	Per cent	No	Per cent
1.	Fully supported	65	65.00	58	58.00
2.	Somewhat supported	25	25.00	26	26.00
3.	Least support	10	10.00	16	16.00

The clients system which actually faced the hardships and constraints while deriving the benefits of livelihood sustainability through agro-biodiversity conservation through the set procedures expressed the existence of livelihood support. Similar findings on the existence hardships were also reported by Marimuthu (1998).

Cultural Values Safe Guarded Agro-biodiversity

As a result of the agro-biodiversity conservation over a period of time, the tribal women would have reaped the benefits of cultural values. Hence it was tested whether the respondents could obtain the benefits of cultural values due to their participation in agro-biodiversity conservation practices that improved their standard of living. The data are presented in Table 4.22.

Table 4.22. Cultural Values Safeguarded Agro-biodiversity.

Sl.No.	Cultural Values	Kothar (n=100)		Irular (n=100)	
		No	Per cent	No	Per cent
1.	Best supported	52	52.00	55	55.00
2.	Somewhat supported	25	25.00	30	30.00
3.	Least supported	23	23.00	15	15.00

As discussed above the cultural values of the community also had been changing in the recent years. The traditional cultural values had the viability to safeguard agro-biodiversity. To what extent it supported the diversity and how it had been perceived by the respondents was the theme of this analysis. The results are depicted in Table 4.22. Nearly half (52.00 per cent) of the Kothars perceived that there was best support from the cultural values. Nearly one-fourth (25.00 per cent) of them perceived that there was some what support. Very meager per cent (23.00 per cent) opined that it was least supportive. In the same way it was among the Irulars.

But on an average all the respondents accepted that they could accrue the benefits due to safeguarding of cultural values. It could be inferred that the respondents could not reap the fullest benefits out of the agro-biodiversity conservation practices. Formal steps could be taken to increase the benefits for the respondents. Marimuthu (2001) found the existence of benefits as perceived by the participants. So the findings of the present study corroborated with the findings of the earlier studies.

Government Policy for *ex situ* Conservation

Keeping abreast of the latest know-how in the agro-biodiversity conservation was a prerequisite to cope up with the requirements of present day scientific agriculture. To know, how best the tribal women replenished their knowledge due to their participation in agro-biodiversity conservation practices the government policies were analysed. The expressed opinions from the study group was collected, analysed and presented below.

Table 4.23. Government Policy for *ex situ* Conservation.

Sl.No.	*Government Policy*	*Kothar (n=100)*		*Irular (n=100)*	
		No	*Per cent*	*No*	*Per cent*
1.	Best supports	45	40.00	42	42.00
2.	Somewhat supports	40	15.00	20	20.00
3.	Least supports	15	45.00	38	38.00

Government policies were important for the participation of respondents in agro-biodiversity conservation. The perception of the respondents on the government policies was given in the Table 4.23. Nearly 45.00 per cent of the Kothars and 42.00 per cent of the Irulars perceived that the government policies had supported *ex situ* conservation. Only one-third of both the group had mentioned that the government policies least supported the conservation. Figure 4.5 depicts the perception of the respondents on the support of government policy on *ex situ* conservation.

Government intervention for supporting the conservation practices provided a way for the people to participate in the conservation practices. The findings reported by Bhuvaneshwari (2005) were also on the similar lines of the present study.

4.3. Participation in Agro-biodiversity Conservation

Participation in *in situ* Conservation

Participation of the tribal women in agro-biodiversity conservation was measured with the help of the components and sub components developed for the study. It was measured with the level of participation high participation, medium and low level of participation with a three point continuum. The participation was

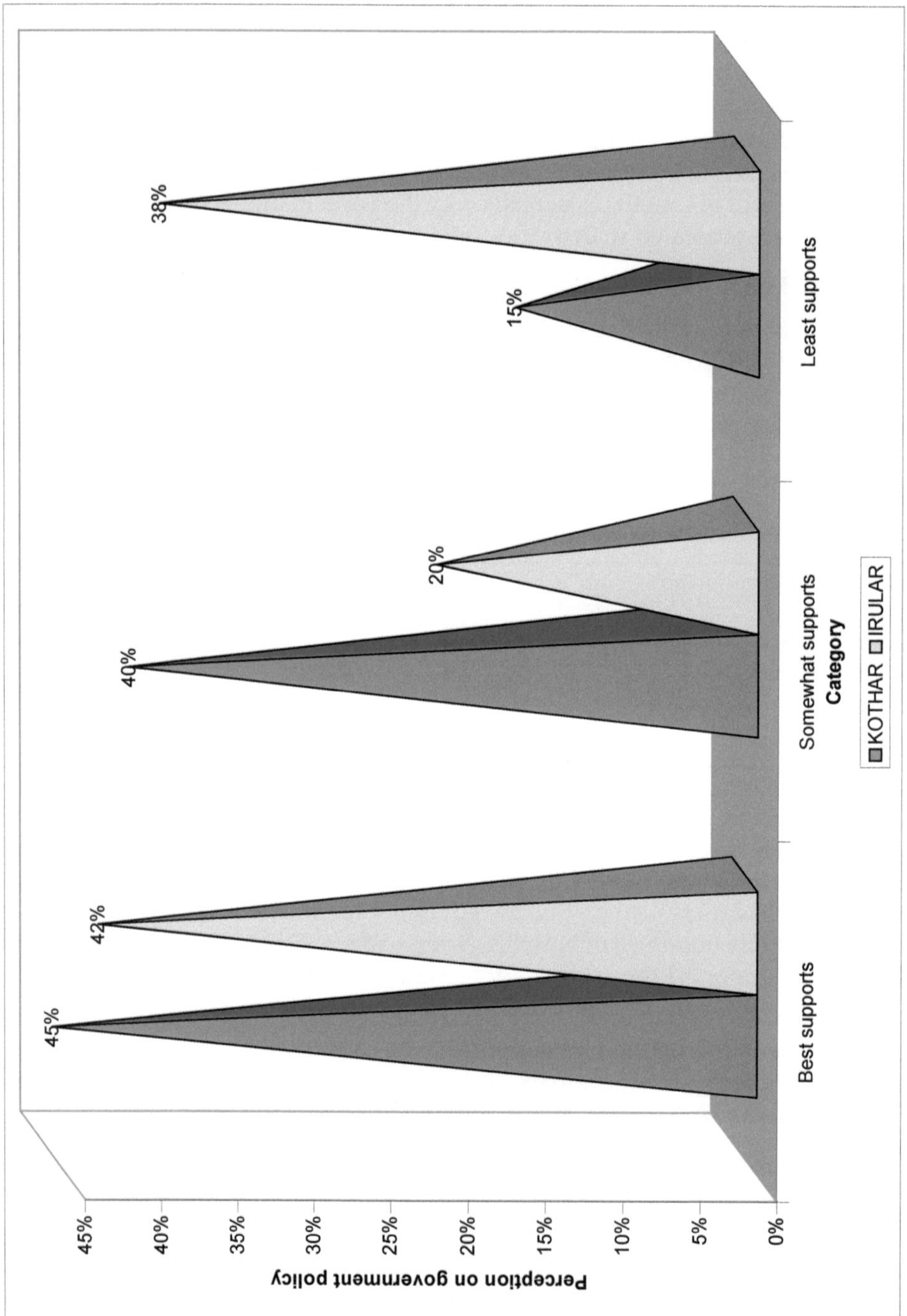

Figure 4.5. Perception on Government Policy for *ex situ* Conservation.

measured as *in situ* and *ex situ* conservation. *In situ* conservation was the agro-biodiversity conservation practices that were done in the field condition. *Ex situ* conservation was the practices that were carried out outside the field condition. In the case of *in situ* conservation the genetic resources and the agro-biodiversity systems played a critical role. The results are depicted in Table 4.24.

Table 4.24. Distribution of Tribal Women according to Participation in *in situ* Conservation.

Sl.No.	*Type*	*Conservation Components*	*Category*	*Kothar (n=100)*		*Irular (n=100)*		*Total (n=200)*	
				No	*Per cent*	*No*	*Per cent*	*No*	*Per cent*
1.	***In situ***	Genetic resources	Low	5	5.00	27	27.00	32	16.00
			Medium	79	79.00	68	68.00	147	73.50
			High	16	16.00	5	5.00	21	10.50
2.		Agro-biodiversity systems	Low	4	4.00	15	15.00	19	9.50
			Medium	22	22.00	6	6.00	28	14.00
			High	74	74.00	79	79.00	153	76.50

Majority of the Kothars (79.00 per cent) had medium level of participation (79.00 per cent) in the conservation of genetic resources followed by high (16.00 per cent) and low (5.00 per cent) levels of participation. Nearly 68.00 per cent of the Irulars had medium level of participation among conservation of genetic resources followed by 27.00 per cent of the respondents with high level of participation. Only 5.00 per cent of the respondents had high level of participation in the conservation of genetic resources. Considering the overall participation three-fourth of the respondents (74.00 per cent) had medium level of participation in the conservation of genetic resources followed by low (16.00 per cent) and high (10.00 per cent) levels of participation. Figure 4.6 explains the distribution of Kothars according to their participation in agro-biodiversity systems.

At the same time the participation of the Kothars in the conservation of agro-biodiversity systems was as follows. Majority of the respondents had high (74.00 per cent) level of participation followed by medium (22.00 per cent) and low (4.00 per cent) level of participation in the conservation of agro-biodiversity systems. Majority (79.00 per cent) of the Irulars had high level of participation followed by low (15.00 per cent) and medium (6.00 per cent) levels of participation. Considering the overall respondents nearly 76.00 per cent of the respondents had high level of participation followed by 14.00 per cent having medium level of participation. Only 10.00 per cent of the respondents had low level of participation in the conservation of agro-biodiversity systems. These findings were in line with Bhuvaneshwari (2005).

Genetic resources were the basic component of conservation of the traditional and wild resources of the crops and the products. Although the conservation of these varieties was too tedious it spoke of the wealth of the system and the community. It was discussed as cultivation of traditional varieties, medicinal plants, and cultivation

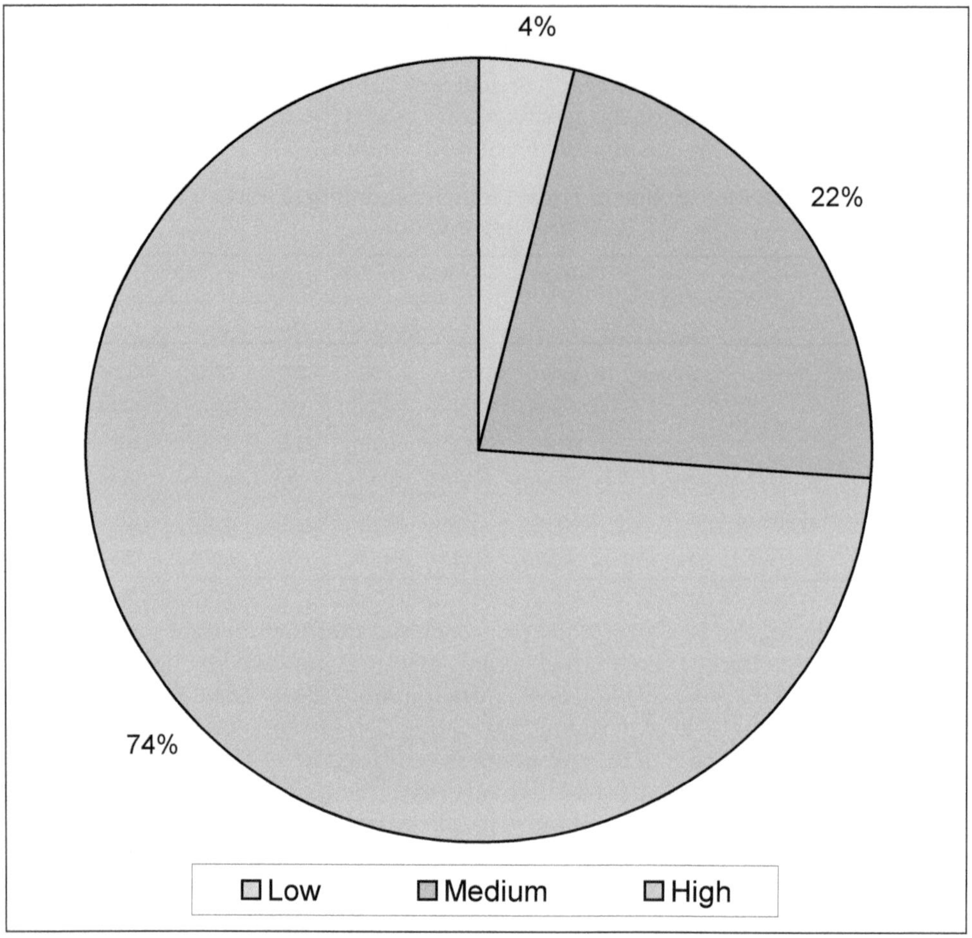

Figure 4.6. Distribution of the Kothars according to their Participation in Conservation of Agro-biodiversity Systems.

of location specific crops like minor millets and domestication of wild varieties as the important aspects of genetic resources. The other important components discussed were collection of forest produce, preservation of forest produce, conservation of soil quality and health, water management practices, livestock maintenance and exotic breeds' husbandry in the case of cattle. These were the important components in the genetic resources for the *in situ* conservation of agro-biodiversity. Considering the conservation of agro-biodiversity systems the important components discussed were water shed management and nutrient management for the crops. Organic cultivation using farm yard manure, adopting natural management practices and utilization of locally available resources were the other important components discussed under this system.

Participation in *ex situ* Agro-biodiversity Conservation

The *ex situ* conservation measures consisted of biotic components and the social values components. Biotic components played a very vital role in the conservation of the agro-biodiversity. This *ex situ* conservation had been measured and the respondents were classified as low, medium and high.

Table 4.25. Distribution of Tribal Women according to Participation in *ex situ* Conservation.

Sl.No.	*Type*	*Conservation Components*	*Category*	*Kothar (n=100)*		*Irular (n=100)*		*Total (n=200)*	
				No	*Per cent*	*No*	*Per cent*	*No*	*Per cent*
1.	***Ex situ***	Biotic components	Low	11	11.00	14	14.00	25	12.50
			Medium	79	79.00	7	7.00	86	43.00
			High	10	10.00	79	79.00	89	44.50
2.		Social values components	Low	4	4.00	18	18.00	22	11.00
			Medium	69	69.00	10	10.00	79	39.50
			High	27	27.00	72	72.00	99	49.50

Among the Kothars 79.00 per cent of the respondents had medium level of participation in the conservation of biotic components, followed by low (11.00 per cent) and high (10.00 per cent) levels of participation in the conservation. Among the Irulars 79.00 per cent of the respondents had high level of participation followed by low (14.00 per cent) and medium (7.00 per cent) levels of participation in the conservation of biotic components. Considering the overall respondents there was an equal distribution between the medium (43.00 per cent) and high (45.00 per cent) level of participation. Nearly 12.00 per cent of the respondents were having low level of participation in the conservation of the biotic components. Figure 4.7 explains the distribution of the Irulars according to their participation in the conservation of biotic components.

Table 4.25. Participation Index of the Respondents.

Sl.No.	*Component*	*Participation Index*	
		Kothar (n=100)	*Irular (n=100)*
	In situ		
1.	Genetic resources	**0.86**	0.84
2.	Agro-biodiversity systems	0.93	0.91
	Ex situ		
3.	Biotic components	0.83	0.85
4.	Social values	0.80	0.83

The other important *ex situ* conservation measure was social values. Two-third of the Kothars (69.00 per cent) had medium level of participation followed by high

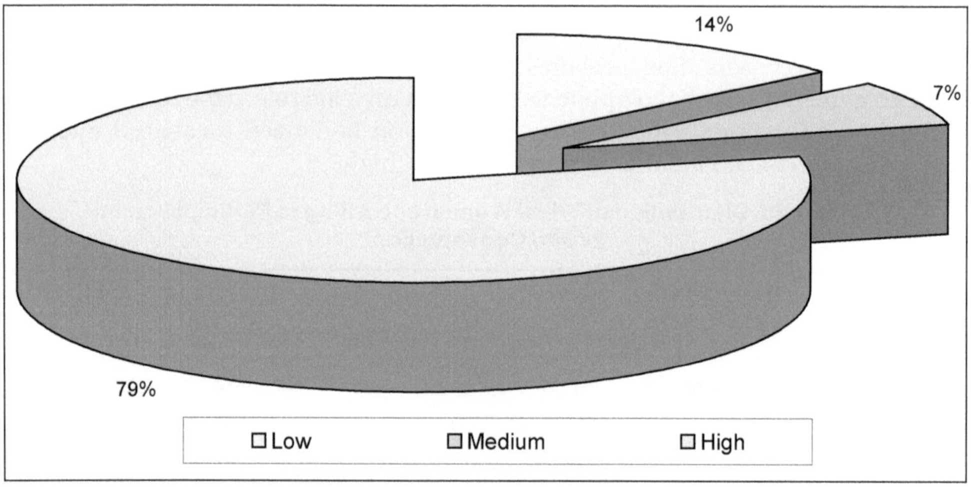

Figure 4.7. Distribution of Irulars according to their Participation in the Conservation of Biotic Components.

(27.00 per cent) and low (4.00 per cent) levels of participation in the conservation of the social values.

Among the Irulars majority (72.00 per cent) of the respondents had high level of participation towards the conservation of the social values, followed by low (18.00 per cent) and medium (10.00 per cent) levels of participation in the conservation Half (50.00 per cent) of the respondents had high level of participation followed by medium (39.00 per cent) and high levels (11.00 per cent) of participation in the conservation.

The component wise participation index was measured and analysed in order to gain a full understanding on the participation of the respondents in agro-biodiversity conservation practices. Among the four components under which the participation was studied, the respondents had high participation in the conservation of agro-biodiversity systems. But the Kothars were somewhat higher than the Irulars in their participation level. Next was the participation in the genetic resources. While considering the *ex situ* conservation components Irulars were superior in conservation activities compared to the Kothars.

Biotic factors and social values were the two important components of *ex situ* conservation of agro-biodiversity. Gene banks, seed banks of traditional crops and medicinal plant gardens were the important components. The other important concepts for discussion were storage of seeds, preservation of extinct varieties, and preservation of the soil biota. The next important *ex situ* conservation practice was social values. Traditional food habits, clothing pattern, cultural behaviour, housing pattern, norms, festivals, indigenous storage practices and post harvest methodologies were the important concepts under the social values. Human intervention in value addition of the crop products, community storage centers, crop management practices and exchanging of the seeds within the community led to measure the level of participation in agro-biodiversity conservation practices. This helped the respondents to maintain a seed bank of various seeds with required qualities. This could be evidently supported by Bhuvaneshwari (2005).

Correlation Analyses between the Profile Characteristics and Participation

Relationship between the Profile Characteristics of Kothar and Participation in *in situ* Conservation

The selection of the independent variables were made based on the discussion with the experts and having high influence towards the dependent variable *i.e.* participation in *in situ* agro-biodiversity conservation. Table 4.27 shows the relationship of the independent variable on the dependant variable on the participation of the tribal women in the *in situ* conservation.

Table 4.27. Correlation between the Profile Characteristics of Kothar and Participation in *in situ* Conservation.

(n=100)

Sl.No.	*Variables*	*'r' Value*
1.	Age	0.088 NS
2.	Educational status	0.106 NS
3.	Farming experience	0.049 NS
4.	Social participation	0.027 NS
5.	Information seeking behaviour	0.331**
6.	Innovativeness	0.095 NS
7.	Economic motivation	0.247**
8.	Progressiveness	0.152*
9.	Cosmopoliteness	0.257**
10.	Attitude towards agro-biodiversity conservation	0.142*
11.	Decision making behaviour	0.242*

**: Significant at 0.01 level; *: Significant at 0.05 level; NS: Non significant.

It was apparent from Table 4.27 that out of the 11 independent variables studied; three variables were having positive and significant relationship at 1.00 per cent level of probability. Three variables were progressiveness, attitude towards agro-biodiversity conservation and decision making behaviour. Apart from these variables information seeking behaviour, economic motivation and cosmopoliteness had significant relationship at 5.00 per cent level of probability. Therefore it could be concluded that the above mentioned six variables alone had significant relationship with the participation of the respondents in the *in situ* conservation.

It was acceptable fact that when the respondents had adequate information seeking behaviour there was likelihood of positive inclination to learn newer ideas, which exposed the family members also to know about the same by the way of discussion and learning through trainings, meetings, exhibitions etc. Hence it was quite natural that information seeking behaviour had positive relationship with the participation of the respondents in the *in situ* conservation.

Economic motivation was another important variable which had positive relationship with the participation of the respondents in the *in situ* conservation. The highly valuable wild varieties and the traditional crops always had some salient features which were always appreciated by the people. The qualities had good market value and fetched a good price for the cultivators. *In situ* conservation led to good economic returns. Hence economic motivation played a positive and significant relationship with the participation of the respondents in the *in situ* conservation.

Progressiveness was the measure of the advanced and scientific thinking of the tribal women. They planned for their future and the next generation ultimately. Scientific measures in conservation approaches were the progressiveness or the progressive thinking of the women. Hence progressiveness had a positive relationship with the participation of the tribal women in *in situ* conservation. Cosmopoliteness was the outward relationship or the information seeking outside their own locality or the community. The communication and the information sharing with the outside communities increased the awareness of the respondents towards the conservation practices. Positive attitude towards agro-biodiversity conservation also had good relationship with the participation of the respondents in *in situ* conservation of agro-biodiversity systems. The decision making of the respondents was ultimately free when there was no any consultation with the other members. This research finding draws supports with the results of Marimuthu (1998) and Bhuvaneshwari (2005).

Relationship between the Profile Characteristics of Irulars and Participation in *in situ* Conservation

It is evident from Table 4.28 that the variables information seeking behaviour, innovativeness, economic motivation and progressiveness had a positive and significant relationship with the participation of the respondents in the *in situ* conservation of agro-biodiversity at 0.01 per cent level. Cosmopoliteness and decision making behaviour had a relationship at 0.05 per cent level. Other variables such as age, educational status, and experience in agro-biodiversity conservation, social participation and attitude towards agro-biodiversity conservation had non significant relationship with the dependent variable.

Information seeking behaviour had positive relationship with the participation in the *in situ* conservation. As information seeking behaviour increased the participation in the conservation practices also increased. The valuable resources were the systems of agro-biodiversity. In order to protect and conserve it the information gathered from various sources were essential. The news given in the TV and radio were important and it spoke about the conservation. Hence information seeking behaviour played a vital role in the participation of the tribal women in the *in situ* conservation.

Innovativeness helped the respondents to adopt a new method of conservation or the technology. High innovativeness always helped the tribal women in obtaining greater knowledge relating to the field. Hence innovativeness helped the respondents to participate more in the conservation practices. Economic motivation also had significant relationship with the participation of the tribal women in the *in situ* conservation. Progressiveness and cosmopoliteness increased the scientific thinking of the respondents. It played a vital role in adoption of the organic cultivation practices, indigenous management practices and the cultivation with locally available resources. Hence progressiveness and cosmopoliteness had positive and significant relationship with the participation of the Irulars in *in situ* conservation.

Table 4.28. Correlation between the Profile Characteristics of Irulars and Participation in *in situ* Conservation.

(n=100)

Sl.No.	*Variables*	*'r' Value*
1.	Age	0.085 NS
2.	Educational status	0.015 NS
3.	Farming experience	0.016 NS
4.	Social participation	0.072 NS
5.	Information seeking behaviour	0.311**
6.	Innovativeness	0.202**
7.	Economic motivation	0.285**
8.	Progressiveness	0.223**
9.	Cosmopoliteness	0.163*
10.	Attitude towards agro-biodiversity conservation	0.131 NS
11.	Decision making behaviour	0.347*

**: Significant at 0.01 level; *: Significant at 0.05 level; NS: Non significant.

Decision making behaviour also decided the thinking and abstract ability of the individual, hence this variable also played a vital role in the participation. As these variables *i.e.* information seeking behaviour, innovativeness, economic motivation, progressiveness and cosmopliteness increased, the participation of the respondents also increased. This result was in accordance with the research evidences of Jamatia (1999) and Bhuvaneshwari (2005).

Relationship between the Profile Characteristics of Kothars and Participation in *ex situ* Conservation

The correlation between the profile characteristics of the respondents with the participation in the *ex situ* conservation is presented in Table 4.29.

Information seeking behaviour and the economic motivation had a positive relationship with the dependent variable at 0.01 levels. Cosmopoliteness and decision making behaviour had a positive relationship with the dependent variable at 0.05 level. The other variables age, educational status, experience, social participation, innovativeness, progressiveness and attitude towards agro-biodiversity conservation had no significant relationship with the participation of the respondents in the *ex situ* conservation.

Table 4.29. Correlation between the Profile Characteristics of Kothars and Participation in *ex situ* Conservation.

(n=100)

Sl.No.	*Variables*	*'r' Value*
1.	Age	0.041 NS
2.	Educational status	0.017 NS
3.	Farming experience	-0.011 NS
4.	Social participation	-0.014 NS
5.	Information seeking behaviour	0.192**
6.	Innovativeness	0.105 NS
7.	Economic motivation	0.206**
8.	Progressiveness	0.118 NS
9.	Cosmopoliteness	0.139*
10.	Attitude towards agro-biodiversity conservation	0.037 NS
11.	Decision making behaviour	0.181*

**: Significant at 0.01 level; *: Significant at 0.05 level; NS: Non significant.

Information seeking behaviour helped the respondents to analyse various sources of information on *ex situ* conservation. Finally it reached a way out such as discussion with relatives, friends and neighbours and provided the conservation measures. Maintenance of gene banks, seed banks and the extract varieties needed a lot of information from various sources. Hence information seeking behaviour played a vital role with the participation of the respondents in the conservation.

Economic motivation was another important variable having positive relationship with the participation of the respondents. As economic motivation increased the participation of the respondents on the conservation also increased. Medicinal plant gardens and kitchen gardening fetched a good income to the respondents. Medicinal plants had a remunerative market and increased the economic condition of the individual. Hence economic motivation had a positive relationship with the participation of the respondents in the *ex situ* conservation. Decision making with individual competency for building up seed banks also influenced the participation of the respondents in the conservation measures. These results were in line with the findings of Bhuvaneshwari (2005).

Relationship between the Profile Characteristics of Irulars and Participation in *ex situ* Conservation

It was observed from Table 4.30 that the variables information seeking behaviour, innovativeness, economic motivation, cosmopoliteness and attitude towards agro-biodiversity conservation had positive relationship with the dependent variable at 0.01 levels. Progressiveness and decision making behaviour had relationship at 0.05 level.

Table 4.30. Correlation between the Profile Characteristics of Irulars and Participation in *ex situ* Conservation.

(n=100)

Sl.No.	*Variables*	*'r' Value*
1.	Age	0.112 NS
2.	Educational status	0.059 NS
3.	Experience in farming	0.021 NS
4.	Social participation	0.066 NS
5.	Information seeking behaviour	0.447**
6.	Innovativeness	0.304**
7.	Economic motivation	0.379**
8.	Progressiveness	0.156*
9.	Cosmopoliteness	0.252**
10.	Attitude towards agro-biodiversity conservation	0.220**
11.	Decision making behaviour	0.365*

**: Significant at 0.01 level; *: Significant at 0.05 level; NS: Non significant.

Information seeking behaviour in the context of the available resources in the community, indigenous management practices, indigenous post harvest practices and value addition had a remarkable contribution. Exchanging the seeds within and outside the community well preserved the components and qualities of the seed material. In this context information seeking behaviour had a positive relationship with the respondent's participation in the *ex situ* conservation. Economic motivation was the other variable that had positive and significant relationship with the dependent variable. The components post harvest practices, value additions, storages and marketing of the produces fetched a good market price for the respondents. This in turn increased the economic returns to the tribal women. Hence economic factors had positive and significant relationship with the participation of the respondents in the socio economic factors. As information seeking behaviour and innovativeness increased the participation also increased.

Innovativeness in the field of the scientific management practices and value addition of the produces would influence the participation of the respondents. Hence innovativeness played a significant role on the participation of the respondents in *ex situ* conservation.

Progressiveness and cosmopoliteness were the other important variables showing relationship with the participation of the Irulars in the *ex situ* conservation. Scientific orientations led to increased progressiveness and cosmopoliteness. Decision making behaviour also had significant relationship with the participation. These results were in line with the findings of Bhuvaneshwari (2005).

Influence of the Profile Characteristics on the Participation

In order to confine the independent variables that influenced the dependent variable stepwise regression was done.

The four variables such as information seeking behaviour, decision making, age and cosmopoliteness had significant contribution on the dependent variable.

Table 4.31 revealed that the R^2 value of 0.575 which showed 57.00 per cent of influence. The F value was 24.09 and found significant. The information seeking behavior alone contributed nearly 48.00 per cent on the participation of the respondents on agro-biodiversity conservation.

Table 4.31. Stepwise Regression Analysis between the Profile Characteristics and Participation

(n=200)

Sl.No.	*Variables*	*Regression Co-efficient*	*Std Error*	*'t' Value*
Step I				
1.	Information seeking behaviour	0.48	0.46	7.696
		$R^2 = 0.480$		
Step II				
1.	Information seeking behaviour	0.38	0.48	5.846
2.	Decision making	0.263	0.38	4.057
		$R^2 = 0.538$		
Step III				
1.	Information seeking behaviour	0.395	0.48	6.147
2.	Decision making	0.254	0.37	3.963
3.	Age	0.15	0.34	2.519
		$R^2 = 0.558$		
Step IV				
1.	Information seeking behaviour	0.373	0.48	5.786**
2.	Decision making	0.128	0.38	3.342**
3.	Age	0.15	0.34	2.541**
4.	Cosmopoliteness	0.146	1.53	2.348**
		$R^2 = 0.575$		

$R^2 = 0.575$ Adjusted $R^2 = 0.317$

F = 24.099 Std error of estimate = 39.59

Along with decision making behaviour it contributed for nearly 53.80 per cent level. Information seeking behaviour, decision making and age influenced for about 55.80 per cent. These variables along with cosmopoliteness contributed to a level of 57.50 per cent. These variables had positive and significant relationship on the participation of the respondents. When there was an increase in these variables there would be an increase on the dependent variable. Figure 4.8 explains the relationship and influence of the independent variables on the participation of the respondents in agro-biodiversity conservation.

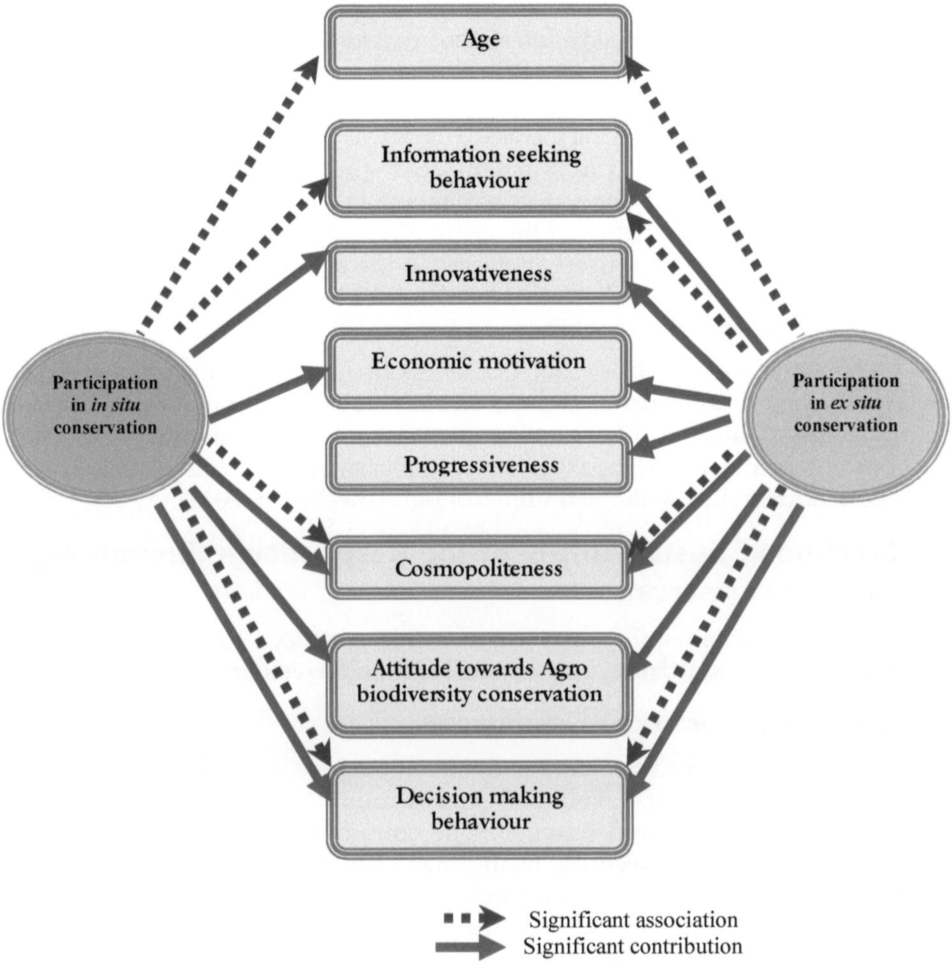

Figure 4.8. Relationship and Influence of the Independent Variables on the Participation in Agro-biodiversity Conservation.

Information seeking behaviour had positive and significant influence on the dependent variable. Information could be gathered from various institutional, non institutional and mass media sources. While there was high information flow from various sources, it increased their knowledge level on agro-biodiversity conservation and ultimately increased the participation of the respondents in the conservation practices. The field experiences revealed that the respondents mainly rely on the non-institutional sources to gather information. They had good rapport with the local NGO's and excellent contact with the extension officials of the department. Mass media also created an impact on the knowledge level of the respondents, which in turn increased the participation of the respondents in agro-biodiversity conservation.

The next crucial variable, was decision making behaviour of the respondents. Participation could be explained as an active involvement of the respondents in the field for conserving the biodiversity. When there was individual decision making it was a boon for the respondents to apply their own ideas in the conservation practices. When this type of decision making increased there was an increase in the participation level of the respondents. Hence information seeking behaviour also had positive and significant influence on the dependent variable. Age was the other crucial variable that influenced the participation of the respondents with the experience they gain through agro-biodiversity conservation. When the age increased it created more conscious among the respondents to conserve the biodiversity. Hence they could concentrate on eco friendly agriculture.

Along with these variables cosmopoliteness also influenced the participation of the respondents. The outside contact apart from their own community was said to be comopoliteness. This type of contacts had increased the awareness and knowledge level of the respondents. Hence their level of participation would also increased. These results are in line with the findings of Bhuvaneshwari (2005).

4.4. Livelihood Sustainability of the Respondents through Agro-biodiversity Conservation

Livelihood sustainability was a systematic approach to link the issues of improvement of standard of living, sustainability and development process.

Relative Weightage of the Components

The next dependent variable was the livelihood sustainability through agro-biodiversity conservation. Livelihood sustainability through agro-biodiversity conservation was measured based on the components identified were food sustainability, nutritional security, health management, shelter security, soil and water sustainability, economic sustainability, employment generation, educational security, assets management, cultural sustainability, infrastructure and marketing sustainability and political sustainability. The relative weightage of each component towards the livelihood sustainability is given in the Table 4.32.

The components having relative weightage scores above the mean score were considered to be the contributing variables. The mean score calculated for the Kothars was 0.93. Food sustainability, nutritional security, health management, shelter security, soil and water sustainability, economic sustainability and employment generation were the components with high relative weightage on the dependent variable. While considering the Irulars, the major components with high relative weightage towards livelihood sustainability were food sustainability, nutritional security, health management, shelter security, soil and water sustainability and economic sustainability. Relative weightage of the food sustainability component towards the Y variable was high while considering the other variables. Food being the basic need of human beings had the high weightage. When the basis diet needs were satisfied it followed with the major supplement of nutritional security.

Table 4.32. Relative Weightage of the Components towards Livelihood Sustainability.

Sl.No.	Components	Relative Weightage	
		Kothar (n = 100) (Mean score 0.93)	Irular (n = 100) (Mean score 0.92)
1.	Food sustainability	0.99	0.99
2.	Nutritional security	0.97	0.95
3.	Health management	0.95	0.95
4.	Shelter security	0.94	0.93
5.	Soil and water sustainability	0.93	0.92
6.	Economic sustainability	0.93	0.92
7.	Employment generation	0.93	0.91
8.	Educational security	0.90	0.91
9.	Assets management	0.90	0.91
10.	Cultural sustainability	0.90	0.90
11.	Infrastructure and marketing sustainability	0.88	0.89
12.	Political sustainability	0.88	0.89

Health management was the next important component towards livelihood sustainability. The rural communities were deprived of the healthy diet one who required which leads to malnutrition and diseases. To lead a sustained livelihood, healthy food and medicinal requirement from the biodiversity was essential. Relative weightage of the components towards livelihood sustainability is given in Figure 4.9.

The field experiences showed that the tribal communities led a life more close to the environment. Majority of them used the locally available materials for their livelihood especially shelter. The houses were thatched and the walls were done with the help of the trees and mini shrubs. Employment generation and economic sustainability go hand in hand when compared to the other components. Their basic work was working in the field as agricultural labours and collecting forest produces. The relative weightage of the sub-components on the livelihood sustainability of the Kothars were comparatively higher than the Irulars. The other sub-components such as educational security, assets management, cultural and political sustainability had least relative weightage on the livelihood sustainability. These results coincided with the results of Marimuthu (2001).

Influence of the Components towards Livelihood Sustainability

The components that had influence on the livelihood sustainability through agro-biodiversity conservation were measured with the factor analysis. The direct and indirect influence of the components on the Y variable could be measured with factor analysis. The results are depicted in Table 4.33. It was revealed that among the twelve components studied seven components influence the livelihood sustainability. These components were food sustainability, nutritional security, economic sustainability and employment generation. The factor analysis showed that these variables had direct influence on the livelihood sustainability.

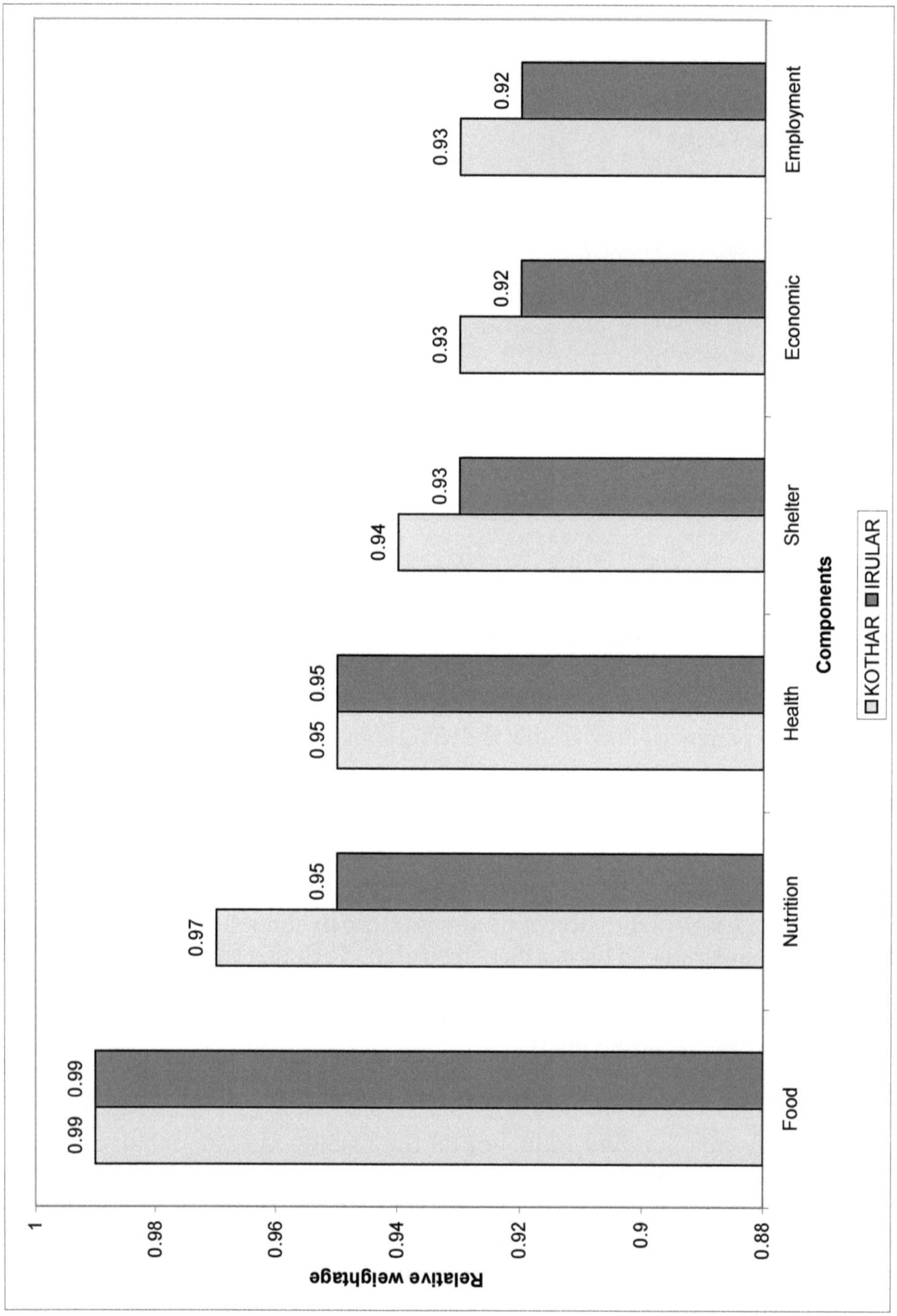

Figure 4.9. Relative Weightage of the Components towards Livelihood Sustainability.

The factor value of food sustainability on the livelihood sustainability was 0.745. This had a direct influence on the Y variable. Food being the basic need of the human was found to be the vital component with high relative weightage and influence on the livelihood sustainability. The socio economic variables of the respondents also had influence on the food sustainability. All the food needs had been satisfied by the respondents from the biodiversity. Traditional food habits were also followed by the respondents. Cultivation of minor millets like varagu, samai and cumbu also sustained the food needs of the respondents. These minor millets played the important part of food needs among the tribals.

Table 4.33. Factor Analysis Showing the Influence of the Components on Livelihood Sustainability.

(n=200)

Sl.No.	*Components*	*Factors*			*Communalities*
		Factor 1	*Factor 2*	*Factor 3*	
1.	Food sustainability	0.745*	0.160	0.240	0.641
2.	Nutritional security	0.787*	0.210	0.215	0.720
3.	Health management	0.856*	0.148	0.194	0.818
4.	Shelter security	0.823*	0.336	0.169	0.837
5.	Soil and water sustainability	0.527	0.257	2.405E-02	0.777
6.	Economic sustainability	0.723*	0.517	0.624	0.673
7.	Employment generation	0.816*	0.171	0.226	0.800
8.	Educational security	0.589	0.142	0.260	0.904
9.	Assets management	0.468	0.219	0.137	0.854
10.	Cultural sustainability	0.489	0.145	0.316	0.681
11.	Infrastructure and marketing sustainability	0.488	7.646E-02	0.194	0.767
12.	Political sustainability	0.489	0.286	0.209	0.839
	Eigen value	8.400	2.591	2.384	9.311
	Per cent of total variance explained	66.347	12.425	7.064	80.689

Compared to the Kothars the Irulars community had more influence on the food sustainability because their cultivation had not been shifted towards the scientific practices. But some piece of lands among the Kothars had been shifted to tea plantation. Hence they could rely on the outside market for vegetable needs. Almost all the household cultivated their own food and satisfied their food needs. Hence food sustainability had influence on the livelihood sustainability.

The second important component was the nutritional security. Apart from obtaining full food security through agro-biodiversity conservation it led to nutritional security also. The natural food was always healthy and it supplied the needed nutrients, vitamins and minerals. Minor millets and vegetables satisfied the requirement of the basic supply of nutrients. Kitchen gardening was practiced in almost all the households during the field visits. These kitchen gardens consist of greens which were known for the fibre and vitamins, vegetables for their unique colours *i.e.* carrot and beet root. Medicinal plants also were found in an important place in the kitchen gardens. Hence the nutrient security had direct influence on the livelihood sustainability of the respondents.

When the food and nutrition status was satisfied it led to the health management aspect. The clear nature and clean environment created a disease free community. Most of the diseases that were identified here were treated with the natural products. It was also discussed that the medicinal plants found an important place among the kitchen and community gardens. The regular phenomenon of each and every rural and tribal community among the Indian villages was the malnutrition and diseases. But these villages were exceptional in these cases. During the field visit it was rare to document such a symptom in these communities. Hence health management had direct influence on the livelihood sustainability of the respondents.

Shelter security was the next important component that had influence on the livelihood sustainability of the respondents. It showed direct influence. Even though there were modern tools and equipments among them the traditional way of building houses was very peculiar among them. Especially the housing style of Kothars was very unique with natural fencing around the house and thatched roofing. The trees of the forests safeguarded the village from natural calamities. The main shrubs were very prominent in these villages. In some cases they had medicinal value also. The Irulars also lived in a natural dwelling of tree leaves and dried shrubs. Every house had a shrub or a live fence. Relationship and influence of the components on livelihood sustainability is depicted in Figure 4.10.

Employment and economic sustainability were the other important components having direct influence on the livelihood sustainability of the respondents. Employment was the base for all the basic needs. Employed respondents would be better citizens and also a boon for their family. This monetary returns and benefits led to the economic sustainability of the respondents. The basic cause for all the problems among the rural and the tribals were the economic instability. So employment would provide a way for them regarding the agro-biodiversity conservation. In turns it satisfied the monetary needs and created an economic sustainability among the respondents. Hence employment and economic component go hand in hand in influencing the livelihood sustainability of the respondents. These discussions coincided with the research findings of Marimuthu (2001).

The empirical model of the study is depicted in Figure 4.11. It was revealed that there was an inter relationship between the participation of the respondents in agro-biodiversity conservation and livelihood sustainability through agro-biodiversity conservation. The crucial variables influencing the participation of the respondents were age, social participation, information seeking behaviour, innovativeness, cosmopoliteness and decision making behaviour. The participation of respondents in *in situ* conservation was higher than the participation in *ex situ* conservation. Food sustainability, nutritional security, health management, shelter security, economic

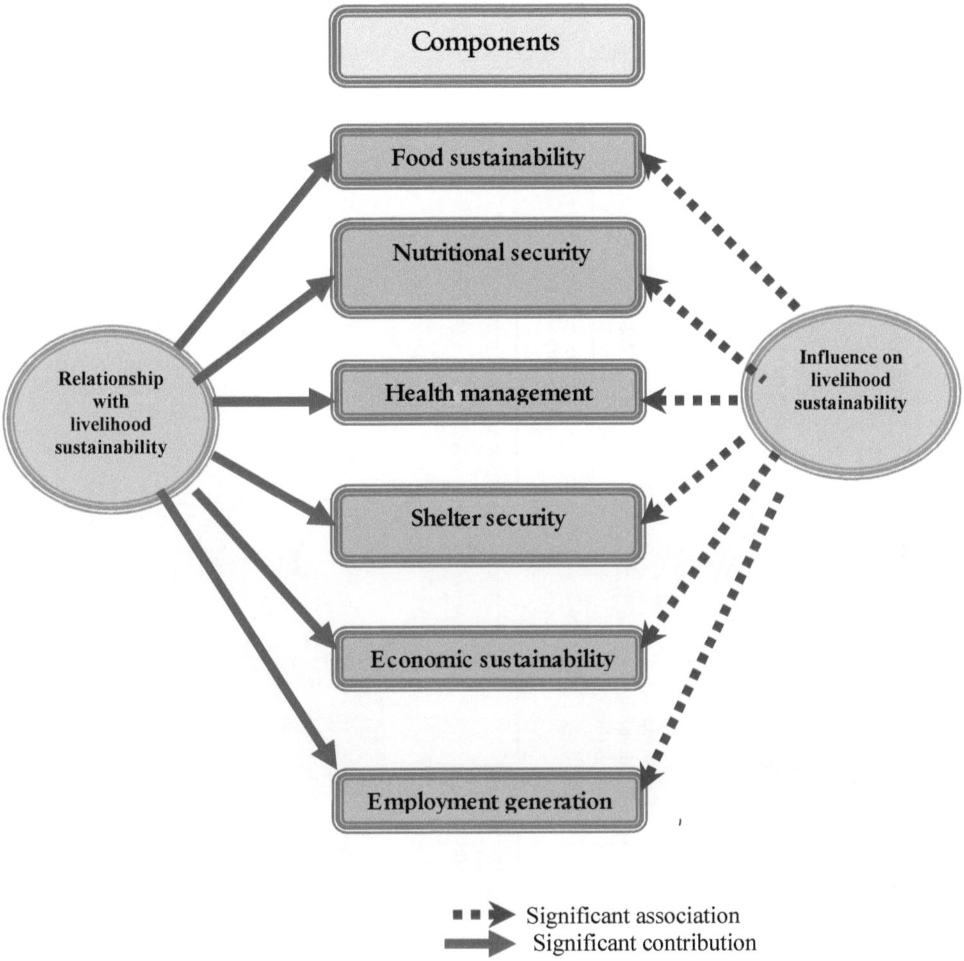

Figure 4.10. Relationship and Influence of the Components on Livelihood Sustainability.

sustainability and employment generation were the important components influencing the livelihood sustainability of the respondents.

4.5. Constraints Faced by the Tribal Women in Participating in the Agro-biodiversity Conservation Practices

The constraints were gathered in the following subheads and it was pooled as multiple responses the results were discussed in percentage analysis. The constraints were classified in to technological constraints, transfer of technology constraints, physical constraints, economic constraints, personal constraints and marketing constraints. Table 4.34 explains the constraints faced by the tribal women in participating in the agro-biodiversity conservation practices.

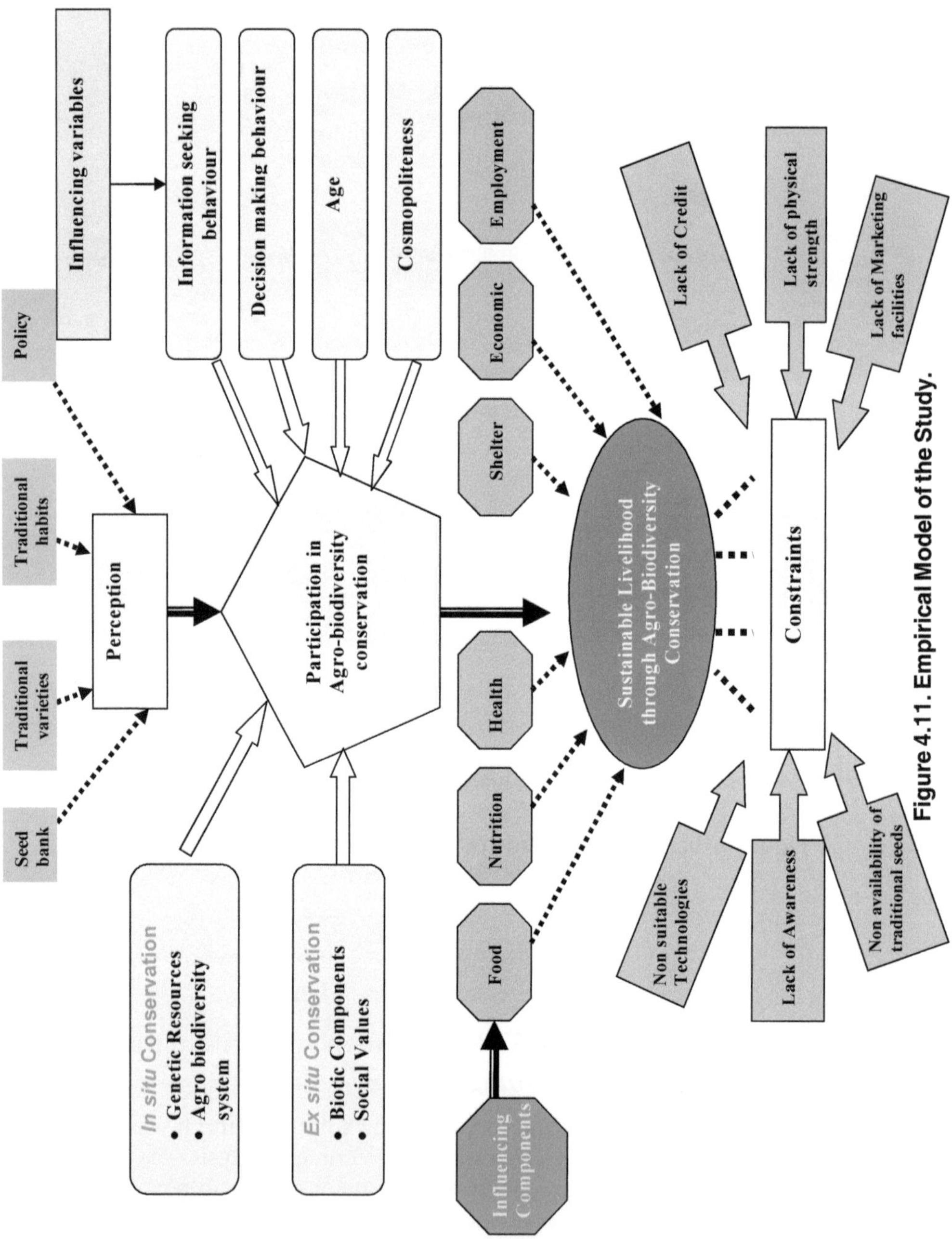

Figure 4.11. Empirical Model of the Study.

Table 4.34. Constraints Faced by the Tribal Women in Participating in the Agro-biodiversity Conservation Practices

(n=200)*

Sl.No.	*Statements*	*No*	*Per cent*	*Ranking*
a.	**Technological constraints**			
1.	Non suitability of the technologies in the recent years	200	100.00	**I**
2.	High risk and uncertainty in adoption of traditional crops	197	98.50	**II**
3.	Complexity in practice	195	97.50	**III**
4.	Lack of easily operating tools to conserve agro-biodiversity	195	97.50	**IV**
b.	**Extension constraints**			
1.	Lack of awareness of recent technologies	200	100.00	**I**
2.	Inadequate technical knowledge in agro-biodiversity conservation practices	195	97.50	**II**
3.	Lack of training	180	90.00	**III**
4.	Lack of upgradation of knowledge in the agro-biodiversity conservation practices	159	79.50	**IV**
5.	Weak extension service	150	75.00	**V**
c.	**Physical constraints**			
1.	Non availability of inputs/traditional seed varieties	200	100.00	**I**
2.	Scarcity of labour	198	99.00	**II**
3.	Shift in working conditions	196	98.00	**III**
4.	Change in the climatic conditions	195	97.50	**IV**
5.	Unfavuorable climatic factors	190	95.00	**V**
6.	Due to migration	180	90.00	**VI**
7.	Non availability of storage facilities	170	85.00	**VII**
8.	Change in soil conditions	160	80.00	**VIII**
d.	**Economic constraints**			
1.	Lack of credit to adopt the agro-biodiversity conservation Practices	200	100.00	**I**
2.	High cost of labour	195	97.50	**II**
3.	High cost of plant protection chemicals	183	91.50	**III**
4.	Delayed cash payment by middleman	160	80.00	**IV**
e.	**Personal constraints**			
1.	Requirement of more physical strength to adopt the agro-biodiversity conservation practices	200	100.00	**I**
2.	Excessive burden of work	190	95.00	**II**
3.	Family norms	183	91.50	**III**
4.	Illiteracy	165	82.50	**IV**

Contd...

Table 4.34–*Contd...*

Sl.No.	*Statements*	*No*	*Per cent*	*Ranking*
f.	**Marketing constraints**			
1.	No market for traditional crops and location specific crops and varieties	200	100.00	**I**
2.	Fluctuation in market prices	195	97.50	**II**
3.	Exploitation by middlemen	193	96.50	**III**
4.	Competition is high in the local market	184	92.50	**IV**
5.	Lack of marketing facility	150	75.00	**V**
6.	Lack of marketing experience	120	60.00	**VI**

*: Multiple responses.

Among the technological constraints, nearly all the respondents perceived that non-suitability of the technologies in the recent years was the major constraint and it ranked first, followed by high risk and uncertainty of the technologies as perceived by 98.50 per cent of the tribal women. Nearly 97.50 per cent of the respondents opined that complexity in practice along with lack of easily operating tools was a constraint. Lack of awareness of recent technologies was the transfer of technology constraint stated by all the respondents followed by inadequate technical knowledge mentioned by 97.50 per cent of the respondents. Nearly 90.00 per cent of the respondents mentioned that lack of training was a constraint. Less extension services and lack of upgradation of knowledge in the agro-biodiversity conservation practices were the other important constraint on transfer of technology. Non-availability of timely inputs was the first and foremost physical constraint among the respondents, followed by the scarcity of labour as mentioned by 99.00 per cent of the respondents. The third constraint was shift in working conditions reported by 98.00 per cent of the respondents. The other important constraints were the change in climatic conditions. Non-availability of storage facilities also was mentioned by nearly 85.00 per cent of the respondents.

Lack of credit to adopt the agro-biodiversity conservation practices was the foremost economic constraint as mentioned by all the respondents followed by high cost of labour. Most of the respondents had mentioned that high cost of plant protection chemicals was the major constraint followed by delayed cash payment by the middleman. Personal constraint played a vital role in adopting the agro-biodiversity conservation practices. Most of the respondents had mentioned that requirement of more physical strength to adopt the agro-biodiversity conservation practices, followed by excessive burden of work, illiteracy and family norms. Illiteracy and family norms were the important constraints that impede the general outlook of the community and working efficiency of the tribal women. Hence their participation in the conservation of agro-biodiversity conservation practices got reduced.

Among the marketing constraints given by the respondents there was no market for traditional crops and location specific crops were the major constraint. For promotion of any technology marketing was very essential. Most of the respondents had mentioned that the high fluctuation in market price was a constraint.

There should be a regularized system to control the fluctuation in the price. It was highly needed for withstanding in agriculture. Exploitation of middlemen and competition in the local market was the other important constraint. When there would be direct marketing from the producer to the consumer the margin could be minimized and the producer could get a remunerative price. Lack of marketing experience was a constraint expressed by nearly 60.00 per cent of the respondents. Many of the producers do not go to the market for direct sale. Hence there was lack in awareness of the exact situation in the market. This unawareness had been exploited by many middle men. Hence marketing experience was must for the tribal women to sustain their livelihood through agro-biodiversity conservation. These were the constraints also expressed by Marimuthu (2001) and Bhuvaneshwari (2005).

4.6. Suggestions for Sustainable Livelihood through Agro-biodiversity Conservation

The suggestions to overcome the constraints in participating in the agro-biodiversity conservation practices were collected based on a closed question format. The suggestions given by the respondents were noted down and it was given as percentage ranking in multiple responses. The data is produced in Table 4.35. There were nearly twenty three suggestions among which providing timely environmental education was the foremost suggestion given by all the respondents. Hence awareness campaigns and exhibitions could be arranged to impart the importance of agro-biodiversity conservation. Training on recent technologies of agro-biodiversity conservation practices and knowledge on technological management were the other important suggestions.

Table 4.35. Suggestions for Sustainable Livelihood through Agro-biodiversity Conservation.

(n=200)*

Sl.No.	*Statements*	*No*	*Per cent*
1.	Providing timely environmental education	200	100.00
2.	Encouraging establishment of seed banks	200	100.00
3.	Providing training on recent technologies of agro-biodiversity conservation practices	192	96.00
4.	Imparting adequate knowledge on technological management	187	93.50
5.	Exchanging knowledge by conducting meeting among groups	170	85.00
6.	Ensuring technical guidance through efficient extension services	167	83.50
7.	Providing inputs/traditional and location specific crops at subsidized rates	162	81.00
8.	Fixing reasonable prices for produces of traditional crops	155	77.50
9.	Developing tools and implements which can reduce the drudgery of tribal women in adopting agro-biodiversity conservation practices	151	75.50
10.	Periodical visits by extension functionaries	143	71.50
11.	Creating awareness on environmental protection	135	67.50

Contd...

Table 4.35–*Contd...*

Sl.No.	Statements	No	Per cent
12.	Recruiting sufficient number of women extension personnel	130	65.00
13.	Utilizing locally available resources	129	64.50
14.	Family labour engagement for agro-biodiversity conservation practices	120	60.00
15.	Developing skilled labour through frequent trainings.	115	57.50
16.	Arranging credit facility for cultivating traditional crops	111	55.50
17.	Arranging marketing facility for traditional and location specific crops	100	50.00
18.	Establishing small processing industries for seeds and produces	95	47.50
19.	Weather and market forecast	94	47.00
20.	Encouraging indigenous practices	92	46.00
21.	Providing infrastructure for medicinal gardening	90	45.00
22.	Providing community storage centers	80	40.00
23.	Policy for traditional crop cultivating system	76	38.00

*: Multiple responses.

Nearly 85.00 per cent of the respondents had given a suggestion about exchanging of knowledge between the other groups by conducting interactive meeting. This could be a useful measure where the ideas could be exchanged between other tribal groups. When there were adequate and timely inputs supplied to the respondents, the constraints could be easily managed. Hence 81.00 per cent of the respondents had suggested about the timely supply of inputs. The other idea was about fixing reasonable prices for the produces. When there were easily operating tools and implements the technology could be easily available to the respondents. Majority (75.00 per cent) of the tribal women were in the need of easy operating tools and implements that could reduce the drudgery of tribal women in adopting the agro-biodiversity conservation practices. Extension personnel play a very vital role in transferring the important technologies on agro-biodiversity conservation practices. Moreover women extension functionaries were having the ability to convince the women respondents and make the technologies reach them immediately. Hence nearly 65.00 per cent of the tribal women had mentioned about the recruitment of women extension personnel.

The other suggestions given by the tribal women were utilization of the locally available resources and engagement of the family labour. So that the expense spent on the conservation practices could be managed easily. Skill development of the agricultural labour was also an important suggestion given by the respondents. Credit facility for the cultivation of the traditional crops was also an important suggestion. After cultivation timely marketing facility was a must. Hence, there should be local market for the traditional and location specific crops. Value addition to any product increased its market value and fetched good price. Hence, there is a need of establishing small processing and value addition units on community basis. In the same way there was an immediate need for the establishment of medicinal plant gardens and also community seed storage centers. Seed bank establishment was the one best way, where in the traditional and location specific varieties could be conserved for the future generation.

4.7. Environmental Education Module (EEM) for Sustainable Livelihood through Agro-biodiversity Conservation

a. Identification of Issues

In order to create an environmental education module the important and current issues were identified. These issues were asked for the opinion of the respondents. On the basis that how many opted for it the results were given in multiple responses and ranked. Table 4.36 explains the important issues identified for environmental education. Some of the important agro-biodiversity conservation issues that need environmental education had been mentioned here. Among the entire measures community seed bank establishment was the important target and need of the hour. A community based seed bank on the traditional and location specific crops could be encouraged. Individual motivation along with the government intervention would make this group approach a successful one. The other important agro-biodiversity conservation issues were the medicinal plant gardens, indigenous practices, soil conservation measures, water management, traditional crop varieties, pollution control in the farm, eco-friendly agricultural technologies, organic cultivation, importance of bio-inputs, medicinal plant cultivation, seed storage and processing along with the value addition centers.

Table 4.36. Agro-biodiversity Conservation Issues for Environmental Education.

(n=200)

Sl.No.	*Issues*	*No**	*Per cent*	*Ranking*
1.	Maintenance of seed banks	200	100.00	I
2.	Development of medicinal plant garden	198	99.00	II
3.	Awareness on indigenous practices	196	98.00	III
4.	Soil conservation practices	194	97.00	IV
5.	Water management practices	190	95.00	V
6.	Knowledge on traditional crop varieties	190	95.00	VI
7.	Pollution control in agriculture	188	94.00	VII
8.	Eco-friendly agricultural Technologies	186	93.00	VIII
9.	Seed storage mechanism	168	84.00	IX
10.	Seed processing unit	160	80.00	X
11.	Value addition centre	150	75.00	XI

*: Multiple responses.

Nearly all the respondents mentioned about the establishment of the community seed banks followed by 99.00 per cent of the respondents opined about the medicinal plant garden cultivation. Medicinal plant gardens were innovative measures to maintain the agro-biodiversity and medicinally important crop plants. Majority of the respondents had mentioned about the importance of soil conservation (97.00 per cent) and water management (95.00 per cent) practices. In the same way traditional crop cultivation also had gained importance. Nearly 93.00 per cent of the respondents need education on eco-friendly agriculture crop technologies. Majority of the respondents had mentioned about the educational requirement of seed storage, seed processing and value addition on the traditional crops, which was an important component on the agro-biodiversity conservation. These information were treated to develop a module for livelihood sustainability through agro-biodiversity conservation.

b. Environmental Education Module (EEM)

The suggested module on environmental education for sustainable livelihood through agro-biodiversity conservation is depicted in Figure 4.12. The empirical evidences from the study were taken to develop the module. It would be utilized for the various stakeholders such as school students, college or university and NGOs. The main theme of this module was agro-biodiversity conservation and livelihood sustainability.

Environment education is education through environment, about environment it is both a style and subject matter. If education as style is concerned, it means using environment as a teaching learning aid and as an approach to education. Regarding the subject matter is concerned; it means teaching about the components and constitutions of environment. The teaching for environment is concerned, it means controlling the environment, establishing proper ecological equilibrium which entails proper use and conservation of resources and also involves control of environmental pollution.

Environmental Education Committee (EEC)

There should be a standard committee at the district or the block level to develop the environmental education programmes and implement them. With the participatory mode the needed issues of the particular area may be identified. This committee should have an environmental scientist to identify the recent issues and develop the subject matter. There should be an extension specialist to develop the approach and strategies to create an impact on the target groups. A representative from a local NGO as needed to identify the location specific problems. This team approach would open the gateway for educational development.

Issues of Environmental Education

The empirical evidences of research earthed out the felt issues of environment. These issues were development of community seed banks, establishment of medicinal plant gardens, awareness on indigenous practices in agriculture, traditional crop varieties of major and location specific crops and eco-friendly agriculture technologies.

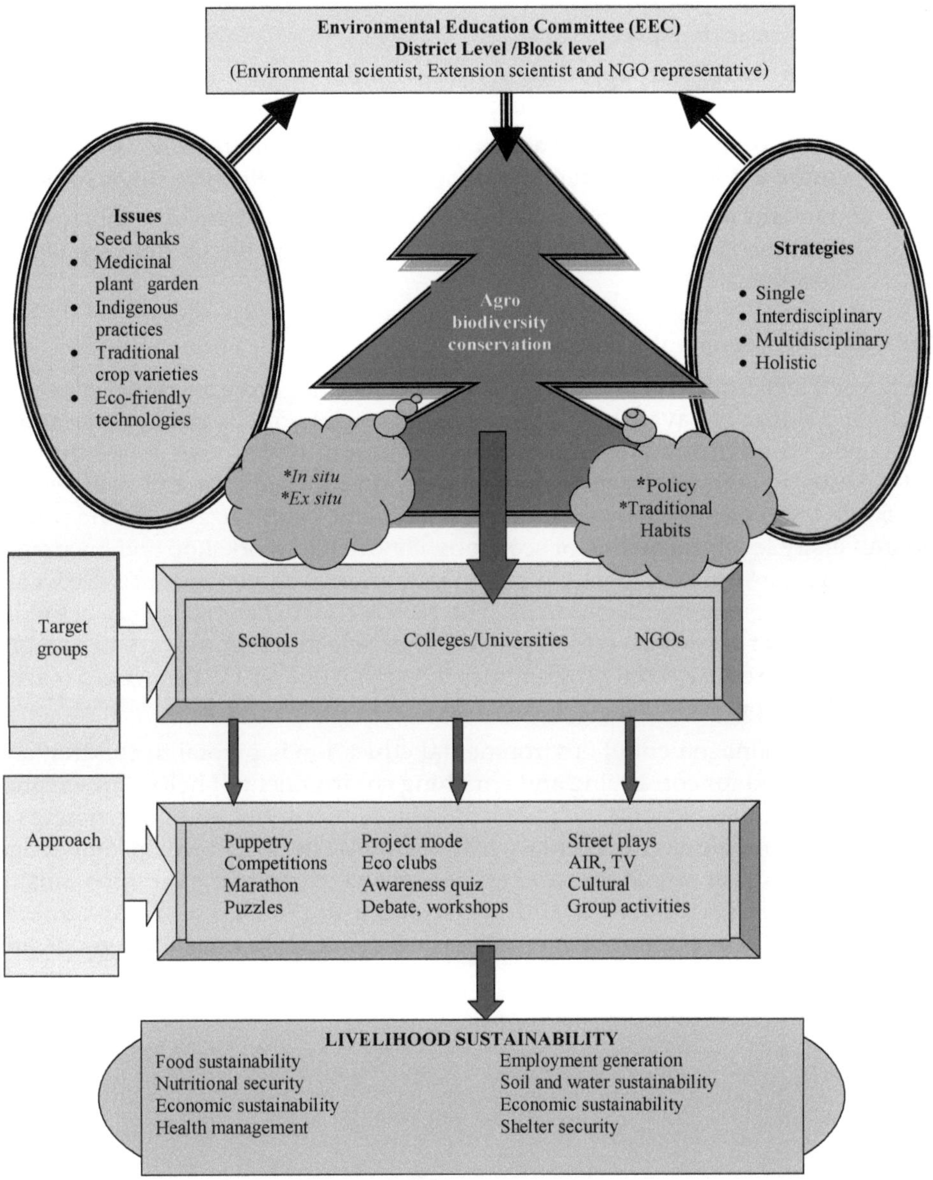

Figure 4.12. Environmental Education Module (EEM) for Sustainable Livelihood through Agro-biodiversity Conservation.

Approaches for Environmental Education

Environmental education should be integrated into the whole system of formal education at all levels. Environmental education curriculum should aim at building up a sense of values. There are four strategies for curricular patterns. They are:

- ☆ **Single subject approach** in which the components are drawn from a single academic discipline
- ☆ **Interdisciplinary approach** in which components are drawn from different discipline rolled into one subject and focus is on single topic
- ☆ **Multidisciplinary approach** in which components are drawn from two or more academic disciplines and focused sequentially on a single topic
- ☆ **Holistic approach** in which there is co-ordination of separate courses such that the diverse fragments of knowledge and understanding are woven together.

Methods of Environmental Education

Several techniques for providing methods of teaching environmental education to school children are available. These includes sending out by mails, entertaining magazines on environment, organization of nature clubs, visit to exhibitions, participation in community environmental action programmes and using songs, plays, folk tales, puppet shows, environmental games and puzzles. At the college and university levels formation of eco-clubs, conducting workshop and awareness quiz would create a great impact. Giving projects on environmental issues also would create awareness among the students. The NGOs could organize group activities with SHGs, street plays and cultural programmes help in creating awareness among the trainees. The educational programmes through AIR and TV also would reach a large number of people.

Effective management of environmental education is pivotal to children and man in their need for conserving and enriching environment. It helped to sensitize individuals, families, schools, neighbourhoods, communities and governments at various levels to solve environmental problems. It also helps to develop knowledge, skills and attitudes required for solving these problems, so that man stops causing havoc by spoiling wild forests and undertaking unplanned and unrestrained industrial activity in pursuit of ever increasing demand for consumer goods and fanciful gadgetry.

Chapter 5

Summary and Conclusion

The world's biological diversity is a vast and undervalued resource. It comprises every form of life from the tiniest microbe to the mightiest beast and the ecosystems of which they are a part. It provides humanity with cornucopia of goods and services, from food, energy and materials to the genes which protect our crops and heal our diseases. Only a fraction of one per cent of the world's total species has been properly studied for its potential values to humanity as food, medicine and industry. People have to be educated about the socio-economic value of biodiversity and that is how our own existence depends upon. Biodiversity would simply mean "varieties of species right to existence". It also would mean that no one has the right to threaten a species right to existence.

The unlimited greed of man is threatening the healthy and diverse biological order that needs immediate correction. It is time to make man realize the negative impact of over-drawing or polluting the nature, *i.e.* 'biodiversity' and make him act as a 'trustee' or 'conservator' of nature in true sense.

To find out suitable and concrete evidences or facts to the above mentioned views, a study was needed. The results of the empirical approaches could be used in making necessary changes in the agro-biodiversity conservation among the tribal women. With this background, the present study was designed and entitled **"Participation of Tribal Women in Agro-biodiversity Conservation and Postulating Environmental Education Module for Sustainable Livelihood"**, with the following:

5.1. Specific Objectives

1. To study the perception of the tribal women on agro-biodiversity conservation.
2. To assess the participation of the tribal women in agro-biodiversity conservation practices.
3. To analyze the livelihood sustainability of the tribal women through agro-biodiversity conservation.
4. To design an environmental education module for sustainable livelihood through agro-biodiversity conservation.

Kothagiri block of The Nilgiris was selected for the study. Kothar and Irulars were the two agriculture tribes selected for the study. A sample size of 200 tribal farm women was fixed for the study. Proportionate sampling technique was followed for the selection of respondents. *Ex-post facto* design was selected as an appropriate research design.

Participation of tribal women in agro-biodiversity conservation and sustainable livelihood through agro-biodiversity conservation were considered as the dependent variables in the present study. Age, educational status, farming experience, social participation, information seeking behaviour, innovativeness, economic motivation, progressiveness, cosmopoliteness, attitude towards agro-biodiversity conservation and decision making behaviour, were considered as independent variables to study their relationship and influence on the dependent variables. Participation of tribal women in agro-biodiversity conservation has been operationalised as the degree to which the tribal women take part in agro-biodiversity conservation through the management and conservation of agro-biodiversity. Genetic resources and agro-biodiversity systems were the *in situ* conservation practices. Biotic components and social values were the *ex situ* conservation practices.

The components that would reflect the sustainable livelihood were identified through various sources such as persuing literature, interaction with the officials of development departments; self help group members, NGO officials and extension personnel. Thus twelve major components were identified for the study. Sustainable Livelihood Index (SLI) was arrived by adding the total scores of each component *viz.*, Food sustainability, nutritional security, health management, shelter security, soil and water sustainability, economic sustainability, employment generation, educational security, assets management, cultural sustainability, infrastructure and marketing sustainability and political sustainability.

Taking into consideration the scope and objectives of the study a comprehensive interview schedule was prepared for data collection. The data gathered were quantified and tabulated for statistical analysis. The statistical techniques applied for the study were percentage analysis, cumulative frequency, correlation analysis, step wise regression analysis and factor analysis. The salient findings of the study are as follows:

5.2. Profile of the Respondents

1. Nearly half (52.00 per cent) of the Kothars were of middle aged group followed by old (30.00 per cent) and young (18.00 per cent) aged group. In the same way nearly half of the Irulars were of middle age group (58.00 per cent) followed by young (24.00 per cent) and old (18.00 per cent) aged group. Among the total respondents 55.00 per cent of them were middle aged group followed by old (24.00 per cent) and young (21.00 per cent) aged group.
2. Less than half of the Kothars were of secondary school education (40.00 per cent) then equally shared by primary (23.00 per cent) and middle (23.00 per cent) school education followed by the illiterate group as 10.00 per cent. Only four per cent of the respondents were of higher secondary education. None were in the collegiate group. Less than half of the Irulars were of primary school education (42.00 per cent) followed by secondary (30.00 per cent) illiterate (15.00 per cent), middle school education (11.00 per cent) and Hr. Sec. (2.00 per cent).
3. Majority (70.00 per cent) of the respondents were having high farming experience followed by medium (23.00 per cent) and low of nearly 7.00 per cent. Nearly half (55.00 per cent) of the respondents were having medium level of social participation followed by 24.00 per cent having high level and 21.00 per cent having low level of social participation.
4. Invariably all the respondents were members in the self help groups in their habitations. It was also noticed that majority of (70.00 per cent) of the Kothars were members in the milk co-operatives and nearly one-third (32.00 per cent) of them were members in the co-op banks. Less than half (44.00 per cent) of the respondents were members in the milk co-operatives followed by 40.00 per cent of them members in the co-operative bank. A meager per cent of Kothars (22.00 per cent) and Irulars (8.00 per cent) were members in the Mahila Mandals.
5. Nearly 74.00 per cent of the respondents were having medium levels of information seeking behaviour followed by 18.00 per cent having high and 8.00 per cent having low level of information seeking behaviour.
6. Nearly three-fourth (75.00 per cent) of the Kothars and 60.00 per cent of the Irulars depend on neighbours followed by family members for information. In the same way 70.00 per cent of the Kothars and half of the (52.00 per cent) Irulars rely on the relatives as their choice of information. In the case of Kothars nearly half of the respondents had opined that officials (63.00 per cent), NGOs (60.00 per cent) and Radio (60.00 per cent) as their sources of information.
7. Nearly two-third (65.00 per cent) of the Kothars were having high level of innovativeness followed by one-third (30.00 per cent) having low level and 5.00 per cent having medium level of innovativeness. Among the Irulars Equal amount of distribution among the medium (46.00 per cent) and high (44.00 per cent) level of economic motivation. Only 10.00 per cent of the

respondents had low level of economic motivation. Nearly half (58.00 per cent) of the overall respondents were having medium level of progressiveness followed by high (26.00 per cent) and low (16.00 per cent) level of progressiveness.

8. Nearly three-fourth (75.00 per cent) of the Kothars was having high level of cosmopoliteness followed by medium (14.00 per cent) and high (11.00 per cent) level of cosmopoliteness. It was evident that high cosmopoliteness prevailed among the Irulars. Among the overall respondents the spread was between somewhat favourable and highly favourable attitude towards agro-biodiversity conservation.
9. Nearly half (52.00 per cent) of the overall respondents had high level of decision making behaviour followed by medium (38.00 per cent) and low (10.00 per cent) levels of decision making behaviour.

5.3. Perception on Agro-biodiversity Conservation Practices

10. Nearly half 45.00 per cent of the Kothars revealed that it was easy to maintain the seed banks in the field. Whereas 32.00 per cent of them had opined that it was somewhat easy and nearly 23.00 per cent of them said that it was difficult to maintain the seeds of traditional crops. Nearly half of both the Kothars (54.00 per cent) and the Irulars (50.00 per cent) opined that the cultivation of traditional crops were easy, only one-fourth of the respondents said that it was somewhat easy and difficult.
11. Majority (60.00 per cent) of the Kothars perceived that *in situ* conservation fully supported their livelihood. In the same way 55.00 per cent of the Irulars also mentioned the same. Nearly one-fourth (27.00 per cent) of the Kothars and 22.00 per cent of the Irulars had mentioned that the *in situ* conservation somewhat supported their livelihood. Only 13.00 per cent of the Kothars and 18.00 per cent of the Irulars felt that *in situ* conservation does not support their livelihood.
12. Majority (73.00 per cent) of the Kothars perceived that there was best improvement in their standard of living. Nearly one-fourth of them perceived that there was some what improvement. Very meager per cent (10.00 per cent and 18.00 per cent) opined that it was least improvement. In the same way it was among the Irulars. Nearly 45.00 per cent of the Kothars and 42.00 per cent of the Irulars perceived that the government policies somewhat supported *in situ* conservation. Only one-third of both the group had mentioned that the government policies best supported the conservation.
13. Nearly 40.00 per cent of the Kothars and Irulars had perceived that the development of seed banks was very difficult. But the same level of respondents had mentioned that it was easy to develop seed banks for *ex situ* conservation. Nearly half (55.00 per cent) of the Kothars and Irulars (52.00 per cent) perceived that traditional food habits best supported *ex situ* conservation. Nearly one-third had mentioned that it somewhat supported.

Only 14.00 per cent of Kothars and 18.00 per cent of the Irulars perceived that it least supported.

14. Majority (65.00 per cent) of the Kothars perceived that *ex situ* conservation fully supported their livelihood. In the same way 58.00 per cent of the Irulars also mentioned the same. Nearly one-fourth (25.00 per cent) of the Kothars and 26.00 per cent of the Irulars had mentioned that the *ex situ* conservation somewhat supported their livelihood. Only 10.00 per cent of the Kothars and 16.00 per cent of the Irulars felt that *ex situ* conservation does not support their livelihood.
15. Nearly half (52.00 per cent) of the Kothars perceived that there was best support from the cultural values for agro-biodiversity conservation. Nearly one-fourth of them perceived that there was some what support. Very meager per cent (23.00 per cent and 15.00 per cent) opined that it was least supportive. Only 40.00 per cent of the Kothars and 42.00 per cent of the Irulars perceived that the government policies supported *ex situ* conservation.

5.4. Participation in Agro-biodiversity Conservation

16. Among the Kothars majority of the respondents had medium level of participation (79.00 per cent) in the *in situ* conservation followed by high (16.00 per cent) and low (5.00 per cent) levels of participation. Among the Irulars 68.00 per cent of them had medium level of participation in *in situ* conservation followed by 27.00 per cent of the respondents with high level of participation.
17. Among the Kothars nearly 79.00 per cent of them had medium level of participation in *ex situ* conservation, followed by low (11.00 per cent) and high (10.00 per cent) level of participation in the *ex situ* conservation. Among the Irulars nearly 79.00 per cent had high level of participation followed by low (14.00 per cent) and medium (7.00 per cent) levels of participation in *ex situ* conservation. Considering the overall respondents there was an equal distribution (43.00 per cent) and (45.00 per cent) among the medium and high level of participation.
18. The respondents highly participated in the conservation of genetic resources. But the Kothars were having higher participation than the Irulars. Next to genetic resources the conservation of agro-biodiversity systems were high among the respondents. While considering the *ex situ* conservation components Kothars were superior in conservation activities while compared to the Irulars.
19. Out of the 11 independent variables studied, progressiveness, attitude and decision making behaviour were having positive and significant relationship at 1.00 per cent level of probability on the participation of the Kothars in conservation. The variables information seeking behaviour, economic motivation and cosmopoliteness had significant relationship at 5.00 per cent level of probability.

20. The variables information seeking behaviour, innovativeness, economic motivation and progressiveness had a positive and significant relationship with the participation of the Irulars in the *in situ* conservation at 0.01 per cent level. Cosmopoliteness and decision making behaviour had a relationship at 0.05 per cent level.
21. Information seeking behaviour and the economic motivation had a positive relationship with the participation of the Kothars in *ex situ* conservation at 0.01 levels. Cosmopoliteness and decision making behaviour had a positive relationship with the dependent variable at 0.05 level.
22. The variables information seeking behaviour, innovativeness, economic motivation, cosmopoliteness and attitude towards agro-biodiversity conservation had positive relationship with the participation of the Irulas in *ex situ* conservation at 0.01 levels. Progressiveness and decision making behaviour had relationship at 0.05 level. The four variables such as information seeking behaviour, decision making, age and cosmopoliteness influenced on the participation of the respondents in agro-biodiversity conservation.

5.5. Livelihood Sustainability through Agro-biodiversity Conservation

23. Food sustainability, nutritional security, health management, shelter security, soil and water sustainability, economic sustainability and employment generation were the components with high relative weightage on the livelihood sustainability of both the Kothars and the Irulas.
24. Among the twelve components studied; food sustainability, nutritional security, health, shelter security, economic sustainability and employment generation were the components influencing livelihood sustainability.

5.6. Constraints Faced by the Tribal Women in Participating in the Agro-biodiversity Conservation Practices

25. Nearly all the respondents perceived that non-suitability of the technologies in the recent years was the major technological constraint followed by high risk and uncertainty of the technologies as perceived by 98.50 per cent of the tribal women.
26. Lack of credit to adopt the agro-biodiversity conservation practices was the foremost economic constraint as mentioned by all the respondents followed by high cost of labour. Nearly 90.00 per cent of the respondents had mentioned that high cost of plant protection chemicals was one of the constraints followed by delayed cash payment by the middleman.
27. The respondents suggested that lack of market for traditional crops and location specific crops were the major marketing constraint. Nearly 97.00 per cent of the respondents had mentioned that the high fluctuation in market price was also a constraint.

5.7. Suggestions for Sustainable Livelihood through Agro-biodiversity Conservation

28. Timely environmental education was the foremost suggestion given by all the respondents. Training on recent technologies of agro-biodiversity conservation practices and knowledge on technological management were the other important suggestions. Nearly 85.00 per cent of the respondents had given a suggestion about sharing of knowledge between the other groups by conducting interactive meeting.
29. Nearly three-fourth (75.00 per cent) of the tribal women was in the need of easy operating tools and implements that could reduce the drudgery of tribal women in adopting the agro-biodiversity conservation practices.
30. Utilization of the locally available resources and engagement of the family labour, skill development of the agricultural labour and credit facility for the cultivation of the traditional crops was also suggested by the tribal women.

5.8. Environmental Education Module (EEM) for Sustainable Livelihood through Agro-biodiversity Conservation

31. In order to create an environmental education module the important and current issues were identified. Among the entire measures community seed bank establishment was the important target and need of the hour. A community based seed bank on the traditional and location specific crops could be encouraged. Individual motivation along with the government intervention would make this group approach a successful one.
32. The other important agro-biodiversity conservation measures were the medicinal plant gardens, indigenous practices, soil conservation measures, water management, traditional crop varieties, pollution control in the farm, ecofriendly agricultural technologies, organic cultivation, importance of bio-inputs, medicinal plant cultivation, seed storage and processing along with the value addition centers.
33. The empirical evidences from the study were taken to develop the module. It would be utilized for the various stakeholders such as school students, college or university levels and NGOs. The main theme of this module was agro-biodiversity conservation and livelihood sustainability.

5.9. Implications of the Study

1. Institutional mechanism like Self Help Groups (SHGs) were fully developed in these tribal villages of the study area. The organized activities of these institutes would tend to build up the values of the society. Hence government intervention is essential to strengthen the dynamics of the SHGs in terms of agro-biodiversity conservation.
2. The study revealed that the information seeking behaviour promoted the efforts to safeguard the biodiversity and the available valuable resources.

Hence appropriate Transfer of Technology (TOT) efforts may be taken up by the local UPASI KVK and NGOs through trainings and other activities for strengthening knowledge on agro-biodiversity.

3. The perception analysis revealed the need for eco-clubs on community participation, wherein storage of seeds, traditional crops and sharing of ideas could be promoted. So creation of newer institutions like eco-hubs in every village would add to the conservation measures. This may be taken up by local bodies duly involving the schools and colleges.
4. Traditional practices, food habits and cultural norms were found to act as factors of biodiversity conservation among tribal women. Hence adequate caution needs to be taken by all the stakeholders not to disturb these traditional values and norms in their endeavour.
5. Overall participation of tribal women in *ex situ* conservation was observed to be low compared to *in situ* conservation. The development department should identify the reasons for such gaps and accordingly execute through out reach programmes.
6. Information seeking behaviour, decision making behaviour, age and cosmopoliteness were found to be influencing the participation of tribal women in agro-biodiversity conservation. Such traits of tribal women have to be considered in selection of target groups for development activities by the policy makers.
7. Agro-biodiversity contributed towards the livelihood sustainability of the respondents. Hence the policy makers may take the stock of livelihood sustainability index on yearly basis and develop appropriate framework for the community.
8. Marketing of the traditional crops emerged as the major constraint among the tribes. Hence the tribal women should get exposed to DEMIC/Agriportal of TNAU for marketing as well as knowledge of marketing procedures and strategies.
9. More of highly favourable attitude towards agro-biodiversity conservation existed among the respondents. So the government could launch an "environmental education module" in the name of nature clubs to benefit the tribal women in order to promote agro-biodiversity conservation measures.

5.10. Suggestions for Future Research

1. Using Sustainable Livelihood Index (SLI), action research projects can be contemplated with various environmental and sociological programmes to study the degree of their participation.
2. Future study may be oriented to test the components in different socio-economic contexts to arrive generalisations on various dimensions of environmental conservation.

3. Similar type of studies can be conducted in various tribal and rural communities.
4. Environmental Education Module (EEM) proposed can be field tested in suitable locale and a generalized frame work can be developed.
5. The validity and reliability of the Environmental Education Module (EEM) can be tested in suitable locations.

Chapter 6

Tools and Techniques Used for Valuation of Biodiversity

***A. Vidhyavathi*[1] *and M. Anjugam*[2]**

[1]*Assistant Professor (Agricultural Economics),* [2]*Professor and Head, Department of Forest Resource Management, Forest College and Research Institute, Mettupalayam – 641 301, Tamil Nadu*

Biodiversity itself, its resources and its functions need to be valued in the context that decline of biodiversity or damage to biodiversity would be very costly to human beings because some of the biodiversity services cannot be replicated. Before going into valuation techniques for biodiversity assessment looking into the basics or principles behind the economic valuation will give an overall idea about how to approach the valuation issues.

In principle, markets will allocate resources efficiently if prices reflect both the full marginal costs of production and the full marginal benefits of consumption, including all components of total economic value. The marginal cost (or benefit) of a good refers to the incremental change in consumer welfare (or producer cost) resulting from an incremental change in the quantity of the good consumed (or supplied). In a competitive market the level of output is determined by prices, which will tend to equilibrate at the point where marginal benefit equals marginal cost and social welfare is maximised. This is the "efficient" level of output, for a given distribution of income. Where prices do not reflect all costs and benefits, however, the so called "invisible hand" of the market does not work and resources may be used inefficiently, resulting in a loss of human welfare. Economists have identified various reasons why market prices fail to reflect environmental costs and benefits. Two of the most important reasons for market failure in biodiversity are the prevalence of "*public goods*" and "*externalities*". *Public goods* are characterized by the fact that: (i) no one can be effectively excluded from consuming them and (ii) increased consumption of the good by one individual does not reduce its availability to others. For example, no one traveling on a public thoroughfare can be charged for enjoying a pretty view, even if the land in question is privately owned. Nor does one person's enjoyment of the view detract from that of another (provided there is no crowding!). Such aesthetic value is among many public goods provided by biodiversity, along with carbon storage and maintenance of ecological balance. Economic theory explains why the free market will systematically under-provide such goods, and why collective action, typically by the government, is usually required to ensure their adequate provision.

Externalities are uncompensated costs or benefits arising from economic activity. A classic example in biodiversity is the decline in availability of game or other non-timber forest products due to logging. Unless the logging company (or landowner) pays compensation to hunters and gatherers for their loss of livelihood, the full economic cost of extracting timber will not have been paid. If similar conditions prevail elsewhere, market prices of timber products will tend to understate true economic costs and consumers will use timber relatively inefficiently.

In addition to public goods and externalities, markets may fail to reflect non-timber forest benefits due to lack of information about their contribution to economic welfare, distortions in prices arising from public policy and regulations, lack of clear or secure property rights over forest lands, and other factors. In such cases, the question arises as to how decision makers can compensate for market failure, and ensure that non-timber forest benefits are given sufficient weight in land use planning and management.

There are many ways to internalise non-market values in the behaviour of producers and consumers, ranging from the introduction of strict environmental standards to ecological tax reform, and from facilitating environmental damage claims in the courts to the promotion of trade in environmental services or "pollution permits".

The aim of valuation is to ensure optimal use of the resources available to society, *i.e.* to allocate these resources where they will be best used and will maximise social well-being. The main advantage of cost-benefit analysis is that it places costs and benefits on a common conceptual basis, reducing them to comparable monetary quantities. This approach is not easily applied in the case of natural assets, however, because many goods and services flowing from the environment are used by human beings in a way that cannot be likened to market consumption. No prices are established for these natural goods. Thus, if such natural resources are to be properly taken into account and allocated in optimal fashion, an economic valuation of these resources must be performed.

Valuation of Biodiversity

While the value of any resource can be assessed by its demand behavior, in case of biodiversity, markets may not exist for all types of products which we obtain from conserving biodiversity while ecosystem 'services' also which provides public goods have no market place at all. Moreover, they have the peculiarity of inter-generational use. As a result of this, standard static economic analysis may not serve the purpose of decision making on issues regarding pricing and distribution. Also, such resources are subjected to a variety of property rights systems, different from individual or private property rights. In such a scenario alternative methods are required to value the biodiversity. There are externalities associated with biodiversity conservation and hence there is a gap between the value and notional price. The positive externalities are generally in terms of various ecological, biological and aesthetic benefits and very little price is paid. Most of the externalities are not accounted for and this results in gross under-estimation of environmental value of biodiversity. Such an under-valuation often leads to inadequate allocation of funds for maintaining the biodiversity hot spots. It is due to these reasons that there is need to take complete assessment of biodiversity and assign economic value to all intangibles including goods and services, soil erosion and agricultural productivity, health, etc.

Points to be Considered before Valuation

Before going for valuation first step to be done is to identify the resources and its services of biodiversity. For example most of the direct and indirect benefits of conserving biodiversity are ecotourism/recreation, education and research, human habitat, watershed benefits, agriculture productivity, soil conservation recharging of ground water, regulation of stream flows,other watershed benefits etc. It includes understanding how these services or functions would change if the ecosystem changed, knowing who are the stakeholders, in what way and for what purpose and what other options they have, and establishing what trade offs exist between different types of services the biodiversity might provide. Once it is finalized the second step is to value of the impacts of the system in monetary terms. To cover all different types of products and services of biodiversity we have to use more than one valuation method. Economic value measures provide a common metric of value for the different services provided by biodiversity *viz.*,carbon sequestration, watershed values, etc. in monetary units, say Rs./hectare. All types of values mentioned above are converted into monetary terms in order to help in decision making on alternative scenarios for the use of biodiversity.

Valuation Techniques

The different techniques can be divided into five broad groups:

1. Market based approach
2. Revealed preference approach
3. Production function approach
4. Cost-based approach
5. Stated preference approach

Market-based Approach

The simplest valuation methods are those, which rely on market prices. Many goods and services from tropical forest land uses are traded, either in local markets or internationally, including wood products (timber, pulp and fuel), non-wood forest products (food, medicine and raw materials). For the products that are commercially traded, market prices can be used to value. The market prices, however, do not always reflect the appropriate prices for the valuation of various forest products. Since many environmental commodities do not pass through the market system, the markets do not provide prices that can be used to value them. For those products that are commercially traded, market prices can be used to construct financial accounts to compare the costs and benefits of alternative forest land use options. In some cases, it may be necessary to adjust market prices to account for market or policy failures. The latter step, also sometimes called "shadow pricing". In addition, there are many services provided by the biodiversity for which there are no prices or it is difficult to have information about their prices. Consequently, a number of other methods can be explored.

Revealed Preference Approach

The basic idea in this approach is that when there are related goods in the market for non-marketed environmental commodity, then, the prices of the related goods may be used. The markets for property/land, labour, and recreational services are used as surrogate markets to infer the value of environmental characteristics of the area/labour market and natural parks, respectively. It is a second group of methods rely on the fact that certain non-market values may be reflected indirectly in consumer expenditure, in the prices of marketed goods and services, or in the level of productivity of certain market activities. These techniques statistically sophisticated methods, such as travel cost models and hedonic pricing, as well as simpler techniques such as the substitute goods method. The theoretical basis for all of these approaches is the household production function, which describes how households attempt to maximise their well-being by allocating time and resources to different activities.

Hedonic Pricing Approach

More specific approaches in this group include the property value approach and the wage labor differential approach, which are used to value environmental amenities and dis-amenities. In both markets, the basic principle is the same: the property (worth) is characterized by several attributes that can be physical, location/ access, and environmental attributes, among other things. Hedonic pricing is based on the assumption that the value of an asset is related to attributes it possesses or the stream of net benefits derived from it. Hedonic pricing is based on the assumption that the market value of land or labour is related to the stream of net benefits derived from it. This stream of net benefits includes a range of factors, including environmental amenities. Therefore, the value of the environmental amenity can be imputed from the observed land or labour market. A particular buyer may be willing to pay for certain environmental attributes in some area that he may not be willing to pay in another area. The value of the property/labor is thus a result of the composite characteristics of the commodity. The specific contribution of the individual attributes can be obtained through a regression analysis to derive the implicit price of say, the environmental attribute (*e.g.*, level of pollution in the area). Once implicit prices are obtained for all the observations, a demand function (an inverse) can be derived for the environmental characteristic and its value derived from an estimation of the area under the demand curve. The market value of a residential property, for example, is affected by many variables including its size, location, construction materials, and also the quality of the surrounding environment. With sufficient data on property values and characteristics it may be possible to control for size, location, construction materials and other factors, such that any residual price differential may be imputed to differences in environmental quality. The hedonic pricing method requires large data sets, in order to account for and eliminate the influence of all other variables, which affect market prices. The approach also assumes that markets for land are competitive, and that both buyers and sellers are fully informed of the environmental amenity or hazard.

Hedonic pricing has been used in developed countries to estimate the negative impact of air and noise pollution, or the presence of waste disposal facilities, on the market prices of residential property and, conversely, the positive impact of proximity to water or public green space. It was found that no examples of hedonic pricing used to assess the environmental amenity value (or any other value) of forests in developing countries. One constraint on use of the technique in developing countries is that private property markets are often thin, uncompetitive and poorly documented. This is a particular problem at the frontier of forested areas, where formal title to land may be missing and where land is often essentially an open access resource.

This situation will probably change as incomes grow and land markets in developing countries become more efficient and discriminating (and as land transactions are better documented). It may already be possible to apply the hedonic pricing method to residential property markets in and around high-growth cities in developing countries, especially where new residential housing developments provide home buyers with the opportunity to reside in greener, forested areas, away from the smog and congestion of the city centre.

Travel Cost Method

This method is used to measure the recreational value of natural resource scenic areas that are normally unpriced or priced with very low entrance fee. The travel cost method is based on the assumption that consumers value a biodiversity site in terms of the cost of getting there, including all direct transport costs as well as the opportunity cost of time spent travelling to the site (*i.e.* foregone earnings). The fee collected is usually based on cost recovery for the staff manning the area and may not even be enough to cover the man-made improvements made in some of the country's national parks. To obtain an idea of how much value people attach to the resource for recreation, the approach assumes that the amount that people are willing to spend to visit the site is a good indicator of the value of the resource to them. This survey-based method has been used extensively to estimate the value of recreational aspects of forests.

Three basic steps are involved in travel cost models. First, it is necessary to undertake a survey of a sample of individuals visiting the site to determine their costs incurred in visiting the site. These costs include travel time, any financial expenditure involved in getting to and from the site, along with entrance (or parking) fees. In addition, information on the place of origin for the journey, and basic socio-economic factors such as income and education of the individual is required. The resulting data is manipulated to derive a demand equation for the site. This relates the number of visits to the site to the costs per visit. The third step is to derive the value of a *change* in environmental conditions. For this, it is necessary to determine how willingness to pay for what the site has to offer alters with changes in the features of the site. By comparing the willingness to pay for sites with different facilities it is possible to determine how the total benefits derived from the site change as the facilities of the site change. A demand function for recreation (measured in visit-days per year or visitation rate per 1,000 population) with price, income, and prices of related goods as dependent variables, should be constructed. From this demand function, the value of the recreation benefits can be estimated.

Substitute Goods Approach

For those biodiversity resources which are non-marketed or which are used directly by the harvester (*e.g.* fuelwood), value may be approximated by the market price of similar goods (*e.g.* fuelwood sold in other areas) or the value of the next best alternative or substitute good (*e.g.* charcoal). The extent to which the value of the alternative marketed good reflects the value of the nonmarket good in question depends, to a large extent, on the degree of similarity or substitution between them. That is, if the two goods are perfect substitutes then their economic values should be very close. As the level of substitution decreases so does the extent to which the value of the marketed good can be taken as an indication of the value of the non-marketed forest good.

Chopra (1993) uses the approach as an alternative to labour cost analysis for valuing fuelwood, in a study of deciduous forests in India. No examples were found of the substitute goods approach being used to estimate other forest values. In some circumstances, a substitute good may be non-marketed as well, but it may still be possible to infer a price for the good in question. In a cost-benefit analysis of a management programme for two forested watersheds in Nepal (Fleming 1983, cited in Dixon *et al.* 1994), fuelwood was valued in terms of the alternative uses of its closest substitute, that is cattle dung which can be dried and burned when wood is unavailable. The opportunity cost of using cattle dung as fuel rather than as fertiliser is estimated in terms of losses in food grain resulting from lower dung inputs into agricultural production. Thus this approach involves two stages: first, estimating a conversion factor for equating the two substitutes (in this case, the energy value of dung and wood); and second, estimating the marginal effect on output, and thus profits, of a change in the use of the substitute (in this case cow dung).

The main problem with the substitute goods approach is that it is difficult to find substitutes which are perfect and which do not provide additional side-benefits. Thus, such value estimates should be viewed with caution.

The Production Function Approach

Another valuation method involves an attempt to relate human well-being or the incremental output of a marketed good or service to a measurable change in the quality or quantity of a natural resource. Such method is variously called production function approach or dose-response or input-output method. The production function approach may be used to estimate the indirect use value of ecological functions of forests, through their contribution to market activities. This approach involves a two-step procedure. The production function approach involves two steps. The first step requires establishing a dose-response function with the response reflecting measures of damages in physical measures (*e.g.*, production losses, incidence of orbidity/mortality). The dose corresponds to the level of environmental stress (*e.g.*, soil erosion, air pollution). The dose response function can be established from data obtained through experiments or survey work.

The second step involves translating the physical impacts to monetary measure. This stage is straightforward to the extent that market prices (adjusted for inefficiency) are available. In the case of loss production for example, the price of the crops may be used in the valuation. Where market prices are not available, cost-based techniques may be employed. To illustrate, morbidity impact may be valued using cost of illness approach. The cost of illness normally consists of hospitalization expenses and work loss days.

An essential requirement of the approach is the availability of reliable information on the physical relationship between the state of the environmental resource and the economic activity or asset it supports. In addition, market conditions and policy distortions affecting production decisions need to be taken into account.

The production function approach has been used extensively in both developed and developing regions to estimate the impacts of changes in environmental quality (*e.g.* deforestation, soil erosion, air and water pollution) on productivity in agriculture, forestry and fisheries, on human health, and on the useful life span or costs of maintaining economic infrastructure.

Cost-based Approaches

The cost-based approaches measure the costs of ensuring the maintenance of the benefits provided by the environmental commodities. One must recognize that these cost measures are not really measures of benefits and should thus be viewed as proxy measures where benefit estimation is not feasible.

The specific methodologies under this category include the opportunity cost method (*i.e.*, the time to collect non-marketed products, usually valued using the wage rate, is used to value the benefits from the products collected). The other approaches are the replacement, restoration, and relocation costs to restore the original condition of the environment. There is also the defensive or preventive expenditures incurred to prevent the change from the existing environmental situation.

In addition to the methods described above for estimating WTP or WTA for non-market forest benefits, some other cost-based approaches may be used to shed light on the costs of maintaining non-market forest benefits, or trade-offs with market values. Three alternative methods focus on the costs of providing, maintaining or restoring environmental goods and services.

The most common methods are:

- Replacement cost methods, which measures environmental values by examining the costs of reproducing the original level of benefits;
- Preventive expenditure methods, which estimate the cost of preventing or defending against degradation of the environment; and
- Opportunity cost approaches, which use estimated production costs as a rough proxy for the value of non-market benefits.

Cost-based techniques are commonly used where there is limited time and resources for more rigorous estimation of environmental benefits. However, such techniques must be used with care, with particular attention to ensure that non-market benefits and costs are not confused. Because cost-based techniques do not directly measure WTP for environmental goods and services, the resulting estimates may over- or under-estimate forest benefits by a large margin. Problems arise when potential rather than actual expenditures are used, as it is not always clear that the environmental benefit in question justifies the costs of replacement, relocation, etc. On the other hand, while cost-based methods are inexact, they may be the only practical alternative in some cases, given resource and time constraints. Where such methods are used, key assumptions about the relationship between estimated costs and associated benefits should be stated clearly.

Replacement Cost Method

The replacement cost technique generates a value for the benefits of an environmental good or service by estimating the cost of replacing the benefits with an alternative good or service. For example, where logging or road construction in upland forest areas leads to increased runoff and sedimentation, some studies use information on the costs of dredging or flood control as a rough estimate of the non-market benefit of watershed protection. The technique rests on the availability of such an alternative, which should - as nearly as possible - produce the same type and level of benefits as supplied by the resource or environmental function being valued. When developing a replacement cost scenario, it is normal practice to select the least cost option among all possible technologies, so as not to over-estimate the value of the environmental benefit.

Relatively few examples of the replacement cost method were found in the literature review. One application to valuing forest benefits is to estimate the value of soil nutrients lost due to increased erosion associated with logging or deforestation in terms of the cost of manufactured fertilizer needed to replace the eroded nutrients.

Preventive Expenditure Method

The preventive expenditure approach (also sometimes called "mitigation" or "defensive" expenditure) places a value on environmental goods and services by estimating the costs of preventing a reduction in the level of those benefits derived from a particular area. This approach may be most applicable for assessing the indirect use values of forests. No case studies using preventive expenditure were found in the literature review. However, a hypothetical example can illustrate how the technique could be used to estimate non-market forest values. For instance, projected expenditures on soil conservation measures aimed at halting or reversing soil degradation could be used as a rough proxy for the benefits generated by the natural nutrient cycling and watershed protection functions of forests. In the case of logging, the watershed protection benefits that might be diminished by building roads for the extraction of timber could be valued by examining the incremental cost of adopting less damaging extraction methods, such as non-mechanised extraction, helicopter logging, or alternative road layouts. As always, it is important to ensure that the benefits of the preventive expenditure match those originally provided by the environmental function, in order to obtain a realistic cost estimate.

The preventive expenditure method is sometimes confused with a variant of the production function approach known as "damage costs avoided". The difference is that the latter approach uses information on the costs of making good or repairing damages incurred as a result of some environmental change, whereas the preventive expenditure approach focuses on the costs of avoiding or mitigating damages before they occur. Where alternative valuation techniques yield different estimates of a particular forest benefit, it is generally preferable to use the lowest estimate so as not to over-estimate the non-market benefit in question.

Opportunity Cost Method

Another valuation approach focuses on the employment / production/ return opportunities foregone in order to secure or protect a particular non-market benefit. As with other cost-based approaches, the focus is on the costs of providing a non-market benefit, rather than the magnitude of the benefit *per se*. The basic idea is that a non-market benefit is worth at least as much as the return that could be obtained by private producers if they were to devote the same effort (*i.e.* the labour used to secure the non-market benefit) in some alternative use.

The opportunity cost approach is most often used to value the subsistence benefits of NTFP collection, where labour is the main input and prices are not available because all or most output is consumed directly by producers. In such cases, the implicit assumption is that a producer's decision to spend time collecting non-timber forest products is weighed against alternative uses of household labour. The opportunity cost of time spent harvesting NTFPs is thus taken as a proxy for the value of the product(s) in question. The only data required are the amount of time spent on the harvest, the resulting yield and the prevailing (rural) wage rate. Nevertheless, care must be exercised when using rural wage rates. First, it is important to ensure that the effort involved in harvesting a non-timber forest product is commensurate with the effort associated with the prevailing wage rate. For example, one day of a child's time spent collecting wild foods should not be valued at the same level as the wage received by an adult for physically demanding agricultural work. In such cases, it may be appropriate to adjust the wage rate used. Similarly, seasonal variations in rural wages should be considered, as the harvesting of NTFPs may be concentrated in periods of slack labour demand, when wages are lower. Third, gathering forest products is often combined with agricultural activities, for example when walking to or from the fields. This must also be taken into account when deciding how much labour time to include in NTFP harvesting costs.

Revealed preference and production function approaches mentioned so far rely on price (revealed preference) to estimate the value of the biodiversity goods and services. These methods try to estimate the value of an environmental good on the assumption that such good exists. An alternative is to ask consumers to state their preferences directly, in terms of hypothetical markets. Here, information on the value of an environmental benefit is obtained by posing direct question to consumers about their willingness to pay for it or, alternatively, their willingness to accept cash compensation for losing the benefit. The most widely used technique in this category is the contingent valuation method (CVM).

Stated Preference Approach

This method is particularly useful when data are unavailable to value the environmental effects of a particular project or when relevant market behavior is not observable. These cases include species preservation; historical or cultural phenomena; scenic, ecological or other characteristics; genetic diversity and preservation of open spaces; unobstructed views or public access to amenity resources.

Contingent Valuation Methodology (CVM)

The CVM technique entails creating a hypothetical market for the good in question and asking the respondents how much they are willing to pay (WTP) for the good in question. The crucial elements of this approach are the definition of the product, the specification of the payment vehicle, and the framing of the WTP questions. The hypothetical nature of the market makes the application of this technique quite open to a number of biases (*e.g.* starting point bias, strategic bias) and should thus be carefully considered. The CVM is applied in such areas as water quality, air quality, biodiversity, and protection of endangered species and land/ recreation facilities studies.

The contingent valuation method uses a direct approach— it basically asks respondents how much they are willing to pay for a benefit, or how much compensation they will be willing to accept to tolerate a cost. The individual's behavior is inferred from the answers he provides in a survey framework. The aim of the contingent valuation method is to elicit valuations which are close to those that will be revealed if an actual market existed. This method is applicable for valuing a wide range of non priced environmental good or service. However, it is more effective when the respondents are familiar with the environmental good or service in question. The success in the use of this method also lies in the design, implementation and interpretation of questionnaires. In biodiversity valuation, CVM has been used to value wildlife and recreational benefits of protected areas. Several recent studies have demonstrated the feasibility of applying CVM to forest land use.

Valuations produced by CVM are "contingent" because value estimates are derived from a hypothetical situation that is presented by the researcher to the respondent. The two main variants of CV are open-ended and dichotomous choice (DC) formats. The former involves letting respondents determine their "bids" freely, while the latter format presents respondents with two alternatives among which they are asked to choose.

Open-ended CVM formats typically generate lower estimates of WTP than DC designs. On the other hand, because no payment is made in most cases, some observers question the validity of stated preference techniques. Critics argue that CVM fails to measure preferences accurately and does not provide useful information for policy. Even practitioners accept that poorly designed or badly implemented CV surveys can influence and distort responses, leading to results that bear little resemblance to the relevant population's true WTP. Much recent attention has focused on overcoming potential sources of bias in CVM studies. Resolving these difficulties involves careful design and pre-testing of questionnaires, rigorous survey administration, and sophisticated econometric analysis to detect and eliminate biased data.

Contingent Ranking

Contingent ranking (CR) is a variant of contingent valuation, and it involves asking respondents to rank a series of alternative non-market goods. One advantage of contingent ranking is that monetary bids may or may not be used. Some have suggested that the use of hypothetical cash payments in CVM may be inappropriate in remote rural communities in the developing world, where people may have relatively little exposure to the market economy. In such cases monetary values can be assigned indirectly, by including in the contingent ranking one or more "anchor" goods with known market values. The contingent ranking method is conceptually simple, easy to administer and able to generate rough estimates of value for a number of forest goods and services at once, without conducting separate WTP surveys for each use and non-use value. On the other hand, contingent ranking may not provide accurate estimates of WTP. A fundamental question is whether the scores generated by contingent ranking reflect an ordinal ranking of relative preference (*i.e.* first, second, third, etc.) as opposed to cardinal measurements of value (*i.e.* one, two, three.). In the former case, it may not be valid to use the cost/price of marketed "anchors" to impute monetary values to non-market goods and services, because they are on different (incommensurate) scales.

Concern in Identification of Stakeholders

The main objective of economic appraisal is to evaluate the costs and benefits of alternative activities in terms of aggregate economic efficiency, *i.e.* irrespective of the distribution of costs and benefits among different groups of people. Most policy-makers, of course, are concerned about distributional impacts, and we may expect alternative forest land use options to impose costs and confer benefits to varying degrees on different social and economic groups.

For example, the designation of biodiversity as protected areas can be seen as a means by which certain interest groups (typically not the poor) secure recreational, amenity or non-use values. This may result in significant loss to another group, *e.g* subsistence farmers who rely on biodiversity land for extraction of non-timber forest products, or for shifting agriculture. Where the values of domestic and foreign consumers differ widely the resulting conflict may be international in scale, as shown by recent heated debates about tropical forestry and timber trade policy in global fora. Similarly, efficiency and equity objectives in biodiversity land use often conflict. Cost benefit analysis may lend weight to certain values and associated land use options that are unavailable to poorer groups, due to their limited access to capital and information. These groups must therefore confine themselves to "inferior" uses. For example, a study of a new national park in Madagascar estimated the value of additional benefits to international tourists at two to three times the loss, in terms of lost agricultural land, incurred by local villagers, despite measures allowing them access to buffer zones (Kramer *et al.* 1995). While such a change in land use may be economically efficient, since the potential for compensation exists, it will aggravate poverty where compensation is absent or inadequate.

Distributional concerns may be incorporated in an economic appraisal of forest land use options in at least three ways: *viz.*, the distributional consequences of land use options can be made explicit by assigning costs and benefits to specific groups, distributional weights can be used in the economic analysis to adjust the benefits and costs according to which groups enjoys or bears them and the entitlements (use and access rights) of particular groups with respect to certain forest resources or benefits may be protected by defining minimum standards or guarantees.

To some extent the first approach is a prerequisite of the second and third. Unless the costs and benefits of a project or land use option can be linked to specific groups there is no way to know where distributional weights should be attached or what rights need to be protected. Therefore, the first step in any distributional analysis is to identify the different *stakeholders* in alternative biodiversity land use options. Depending on the region and the particular land uses in question, this may include indigenous hunter/gatherer populations, subsistence farmers, commercial farmers, small-scale traders, industrial firms (owners and employees), local, state and national government agencies and domestic and foreign consumers.

The next step is to determine which groups are affected by the various impacts of alternative land use options. Some costs and benefits may be spread widely among a number of groups, while in other cases the impact on certain stakeholders will be more concentrated. For example, the benefits of timber harvesting will be spread among the owners of logging companies and their employees, as well as firms involved in providing equipment, wood processing, transport, distribution and sales. It may not be possible to single out every industry (let alone every firm) which benefits from a particular forest land use option. However, it is usually possible to distinguish impacts on broad sectors of the economy and on different categories of labour (*e.g.* skilled versus unskilled).

Finally, the link between costs and benefits and different groups needs to be quantified, if possible, to show the magnitude of the distributional impact. Ideally this will be in monetary necessary to describe their impact on different stakeholders in physical or qualitative terms. Costs and benefits may be expressed in financial terms, using market prices, or in economic terms, with adjustments made to account for market imperfections and/or policy distortions, although it is preferable to use the former when looking at distributional impacts. It may be more difficult to trace the distribution of *non-marketed* costs and benefits, although the techniques used to value these items can often be extended to distinguish different groups. For example, in the case of watershed protection benefits provided by an upland forest, it may be possible to identify those who stand to lose if these services are disrupted, *e.g* land-owners and residents of the downstream flood plain, the regional water or irri irrigation management authority controlling a downstream reservoir subject to sedimentation, etc.

Practical Difficulties of Economic Valuation of Biodiversity

In theory, these valuation methods can be applied to biodiversity in order to calculate its total economic value, *i.e.* its contribution to human well-being. In practice, however, the concept of total economic value seems valid theoretically, in the real world it provides only partial and often subjective information on the benefits that can be expected from use of biodiversity.

Four reasons are generally advanced to explain why this valuation exercise is so difficult:

1. The total economic value of an ecosystem can be estimated directly only by using the contingent valuation method, and this technique is difficult to apply in contexts having a low level of monetisation.
2. Monetary quantification of natural assets is a tricky matter because we have only partial knowledge of how ecosystems work.
3. It is very frequently observed that the assumptions used to estimate economic values are deliberately conservative: given the uncertainty of the valuation procedure, the analyst generally opts for a low estimate of the benefits derived from the environment. This choice indicates how much latitude the analyst has in producing the estimate.
4. The literature shows that, in practice, estimates of total economic value result not from aggregating all the benefits drawn from the ecosystem concerned, but from aggregating only certain values that the analyst was able to quantify in monetary terms. The concept of total economic value thus corresponds to the sum of a few economic values selected subjectively by the analyst, rather than to the sum of all the values that actually constitute total value.

These practical difficulties do not, however, negate the legitimacy of economic valuation of biodiversity. Although issues in estimating monetary valuation are indeed an ideal objective, monetary valuation of certain benefits of biodiversity often provides important information for users and/or managers of these resources.

References

Chopra, K. 1993. "The Value of Non-Timber Forest Products: An Estimation for Tropical Deciduous Forests in India" in *Economic Botany*, 47(3):251-257.

Haripriya, G.S., 2001. Carbon Budget of Indian Forest Ecosystem, Indira Gandhi Institute of Development Research, Mumbai.

Chapter 7

Information and Communication Technology (ICT) for Sustainable Livelihood

V. Anandhi[1] and C. Cinthia Fernandaz[2]

[1]Assistant Professor (Computer Science),
[2]Assistant Professor (Agricultural Extension),
Department of Forest Resource Management,
Forest College and Research Institute,
Mettupalayam – 641 301, Tamil Nadu

Information and communication technologies (ICTs) have, in the recent past, attracted enormous attention as important tools for socio-economic development. The Sustainable Livelihoods approach is a holistic and flexible framework for understanding, measuring, and analyzing poverty and poverty alleviation. Strengthening livelihoods and building resilience among marginalized communities and fostering innovative development. The sustainable livelihoods idea was first introduced by the Brundtland Commission on Environment and Development, and the 1992 United Nations Conference on Environment and Development expanded the concept, advocating for the achievement of sustainable livelihoods as a broad goal for poverty eradication. Sustainable Livelihood is to be applied to enterprise development; specifically to enterprises where information and communication technologies (ICTs) are used promote sustainable livelihoods. A livelihoods approach to ICT-based enterprise should provide insights of particular relevance to issues such as vulnerability, sustainability and impact; especially in relation to those ICT-based enterprises that touch upon the lives of the poor. The approach also facilitates an understanding of the underlying causes of poverty by focusing on the variety of factors, at different levels, that directly or indirectly determine or constrain poor people's access to resources/assets of different kinds, and thus their livelihoods. It also provides a more realistic framework for assessing the direct and indirect effects on people's living conditions than, for example, one dimensional productivity or income criteria.

There are three basic features common to most approaches. The first is that the focus is on the livelihoods of the poor. The second is that the approach rejects the standard procedure of conventional approaches of taking as an entry point a specific sector such as agriculture, water, or health. And finally, the SL approach places great emphasis on involving people in both the identification and the implementation of activities where appropriate. The sustainable livelihoods approach (SLA) is a way to improve understanding of the livelihoods of poor people. It draws on the main factors that affect poor people's livelihoods and the typical relationships between these factors. It can be used in planning new development activities and in assessing the contribution that existing activities have made to sustaining livelihoods.

The two key components of the SLA are:

- ✰ A *framework* that helps in understanding the complexities of poverty
- ✰ A set of *principles* to guide action to address and overcome poverty

SLA has seven guiding principles.

- ✰ ***Be people-centred:*** SLA begins by analysing people's livelihoods and how they change over time. The people themselves actively participate throughout the project cycle.
- ✰ ***Be holistic:*** SLA acknowledges that people adopt many strategies to secure their livelihoods, and that many actors are involved; for example the private sector, ministries, community-based organizations and international organizations.
- ✰ ***Be dynamic:*** SLA seeks to understand the dynamic nature of livelihoods and what influences them.
- ✰ ***Build on strengths:*** SLA builds on people's perceived strengths and opportunities rather than focusing on their problems and needs. It supports existing livelihood strategies.
- ✰ ***Promote micro-macro links:*** SLA examines the influence of policies and institutions on livelihood options and highlights the need for policies to be informed by insights from the local level and by the priorities of the poor.
- ✰ ***Encourage broad partnerships:*** SLA counts on broad partnerships drawing on both the public and private sectors.
- ✰ ***Aim for sustainability:*** Sustainability is important if poverty reduction is to be lasting.

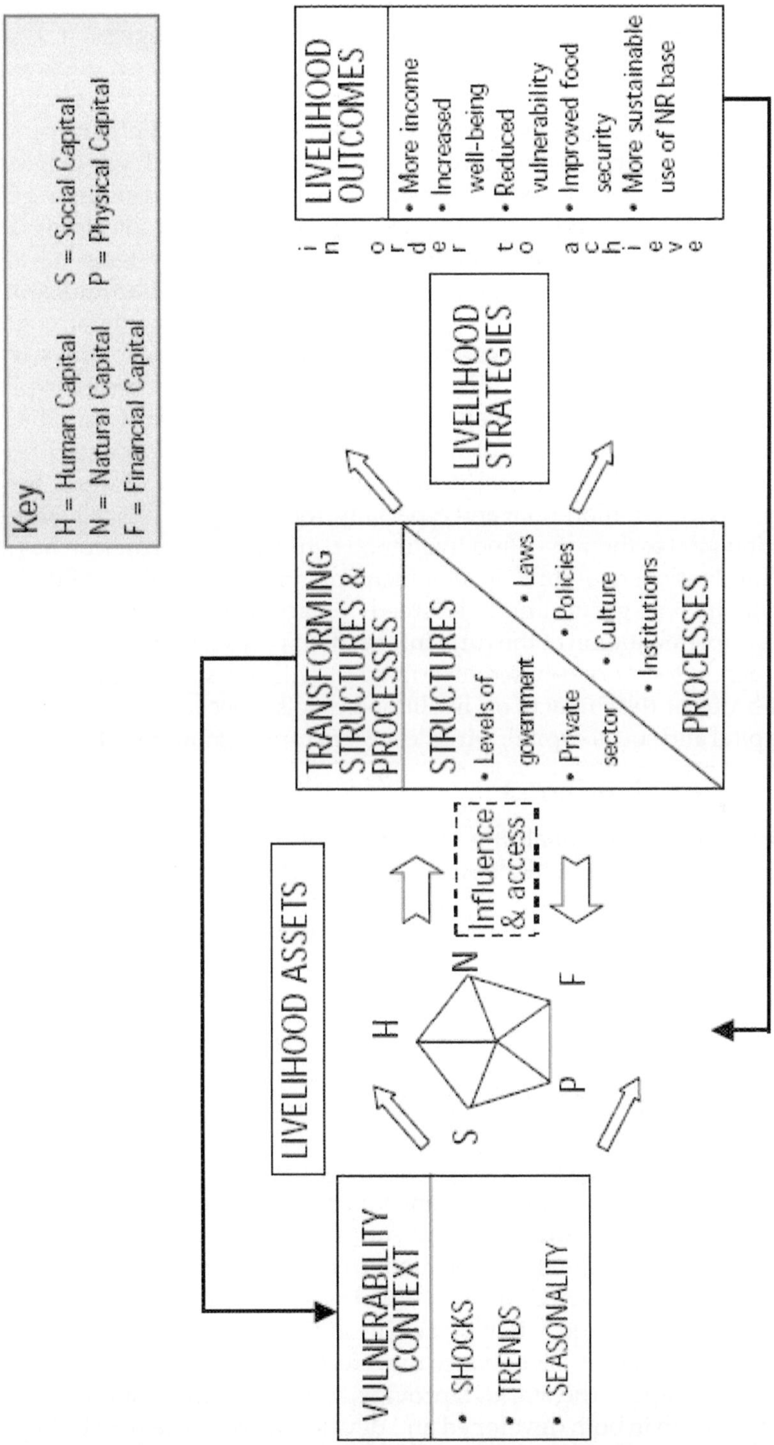

Figure 7.1: Sustainable Livelihood Framework (DFID 2001).

Information and Communications Technologies (ICT)

ICT Stands for "Information and Communication Technologies." ICT refers to technologies that provide access to information through telecommunications. It is similar to Information Technology (IT), but focuses primarily on communication technologies. This includes the Internet, wireless networks, cell phones, and other communication mediums. Information and Communications Technologies (ICT) education is basically our society's efforts to teach its current and emerging citizens valuable knowledge and skills around computing and communications devices, software that operates them, applications that run on them and systems that are built with them. ICTs are technologies offering new ways for communicating and exchanging information and knowledge. They can be used to enable, strengthen or replace existing information systems and networks. The challenge for those working in ICTs is to define the particular roles that information can be expected to play and where ICTs might be most effectively applied but equally to clarify and be honest about what they cannot do. Information and communication activities are a fundamental element of any rural development activity. Rural areas are often characterized as information-poor and especially, the rural poor typically lack access to information vital to their lives and livelihoods. Information provision has always been a central component of rural development initiatives. The vulnerabilities arising from location, income group impinge directly on access to information. However, through the implementation of the rural information project, more rural populations have equal access to internet-based information. This transforming of structure by the project brought the impacts on livelihood assets, specially on human capital, financial capital and social capital, while less on natural capital and physical capital.

Technology for Livelihood

The technologies are enabling citizens to be creative for livelihood. With technology, low-level citizen like farmers are now able to bypass middlemen who used to buy their crops on very low prices by checking the prices on their mobile phones (Rwanda case: eSoko). Farmers are now able to connect to each other and to agronomists and other sponsors via mobile phones and smartphone applications.

One of the most powerful tools for creating development in today's world are the new information and communications technologies that are available in most countries, often at extremely unequal levels. Young people are often the leading innovators in the use and spread of ICTs. Youth are increasingly using these technologies as tools for social change in their lives. They are organizing and creating global initiatives and businesses using these new technologies. The advantages of making these tools available and applicable to youth in countries around the world will have a great impact at addressing the youth employment challenge. Providing technology alone to areas devoid of the modern machines and equipment is not the goal. The aim is to empower Country Networks around the world to use ICTs to spearhead their activities and action plans. The initiative will explore ways in which computers and the Internet can enhance the effectiveness and impact of network in various countries around the world. If provided with an enabling environment and opportunities, youth in both developed and developing countries can be key agents for social change, economic development and technological innovation. Youth bring with them boundless energy, imagination, creativity, ideals, and a limitless vision for their future and the societies in which they live. If not utilized, they are a wasted resource. Thus, it is imperative that youth are harnessed as part of society. This can be achieved through providing sustainable and employment and livelihoods opportunities for them. ICT infrastructure will reshape the urban environment to respond to 21st century challenges. Vibrant and creative cities drive economic, social and cultural development. Governments rightly focus on renovating existing cities and cultivating new ones to act as a catalyst to progress. This is now taking place on an unprecedented scale, with governments and developers launching projects that will change the urban landscape.

- ✰ **Governments have an opportunity to ensure ICT infrastructure is integrated with the changing city:** This will create a Smart City that enhances the lives of citizens, encourages business to invest, and create a sustainable urban environment. Thinking Smart now will ensure that a city retains its competitive edge.
- ✰ **Citizens can benefit from enhancements to the services they use enabled by Smart infrastructure:** Governments need to play a direct role in enabling the Smart City infrastructure and delivering Smart services. Enhancing communications between people and government will enhance the delivery of services creating a safe, friendly and efficient environment.
- ✰ **A more attractive place to do business:** Smart infrastructure also acts as a catalyst for economic development. The availability of high speed, low latency networks with excellent global connectivity enables business to benefit from services that exceed anything available elsewhere.
- ✰ **Supporting and enabling a sustainable future:** Increasingly, a key challenge for cities everywhere is ensuring sustainability, and driving towards the development of a low carbon economy. ICT infrastructure will support better use of resources and more efficient operation of the city.
- ✰ **Smart City infrastructure can create significant economic benefits:** There are strong incentives for implementing the Smart infrastructure beyond introducing innovative smart services. ICT investment boosts economic productivity, contributing between 0.3 per cent -0.7 per cent of GDP per capita growth
- ✰ **What constitutes a Smart City?** The Smart City requires a ubiquitous ultra-high speed network infrastructure, fixed and wireless, that allows people, business and Government to connect with each other and the systems that manage the infrastructure and services of the city.

It was observed that inadequate road infrastructure makes it difficult for farmers to transport their produce to markets outside the districts, even in cases where farmers are provided with information on the availability of those markets. Inadequate electricity supply hinders large-scale uptake of ICTs in rural areas. The findings show that telecentres managers were not aware of the information needs of the communities they serve and therefore they were not in a position to meet such information needs. On the relationship between ICT and rural livelihoods the findings of the study show that while ICTs may not fully support and sustain socio-economic development in poor communities, the impact of these technologies extends to various aspects of the livelihoods of the rural people. Economically, these technologies lead to better earnings and savings. Socially, they help in community interaction and knowledge-sharing, better follow-up for remittances and creation of savings and credit co-operative societies. In relation to human capital, the technologies implications extend to ICT literacy, improved farming techniques and better access to information on new cash crops. The use of ICTs in rural areas is still faced with many barriers. These range from affordability to language problems and the lack of basic infrastructure such as electricity. Language was particularly found to be a substantial barrier to effective use and application of ICTs.

Promoting the Development of the ICT Industry

- Raise awareness of ICT as an educational component
- Raise awareness of ICT as a business component
- Endorse an ecient drive towards the modernization of ICT-related laws and regulations
- Improve transparency and predictability of laws related to ICT

Key ICT Challenges and Opportunities

- Encourage and promote the use of locally suitable software products
- Encourage, liberalize and standardize the use of
- Geographical Information Systems (GIS)
- Encourage, liberalize and standardize the use of Global Positioning Systems (GPS) and remote sensing
- Monitor and control Intellectual Property Right issues
- Develop and implement standards and guidelines for the privacy and condentiality of information
- Develop and implement standards and guidelines for security
- Develop incentives for ICT usage in the private sector
- Raise awareness for the need and use of ICT strategic planning in both the public and private sectors
- Encourage public private partnership for national ICT projects

Improve the Use of ICT in All Sectors

- Promote and implement good ICT practices in the private sector through raising awareness of standards, software development methods, project management and quality management for ICT resources and operations
- Raise awareness of specic ICT usage in various sectors such as agriculture, tourism, trade, etc.

- Use ICTs to provide national and timely business information for healthier business competition
- Encourage the use of consolidated Decision Support Systems (DSS) or Business Intelligence (BI) in the private sector. Dierent sectors can consolidate their information to provide a better assessment of their situation
- Encourage new ICT-based work modes such as telecommuting, on-line jobs, telemarketing, etc.
- Promote the use of ICTs through incentives, exemptions and grants
- Use ICTs to improve business opportunities within and outside Rwanda
- Use ICTs to encourage the establishment of cooperatives for an increased eciency in promoting rural business and industries in urban areas in Rwanda

Improve the Efficiency of the ICT Sector

- Initiate and support research and development in the ICT Sector
- Improve ICT practices in both the private and public sectors
- Encourage fair business practices within the sector
- Develop the proper legal framework and regulations for ICT operations, import and export, and employment contracts
- Ensure on-going collection and dissemination of information about the ICT Sector
- Modernize patent and copyright laws
- Improve public sector procurement procedures for ICT products and services

Assist the ICT Sector in its Objectives

- Enhance the image of Rwanda as a center of ICT knowledge and expertise
- Reward excellence in ICT innovation and use
- Promote the start-up of Application Service Providers (ASP) as this would encourage export of services to regional countries
- Encourage the export of business/ICT expertise through outsourcing
- Provide tax exemption incentives on ICT expenditure in both ICT and other sectors
- Raise awareness of ICT usage and benets in the business culture
- Endorse policies that reduce costs (mostly labour) to facilitate export
- Encourage and facilitate ICT exports
- Encourage and facilitate Foreign Direct Investments (FDI) in ICT
- Encourage and facilitate nancial support for start-up ICT rms (e-grants)
- Encourage research projects aimed at solving local problems and meeting national standards

- ✰ Improve the business capabilities of technological companies
- ✰ Sponsor the presence of ICT companies in regional and international technological exhibitions

The particular focus is on female-led enterprise given the growing and compelling body of evidence suggesting that persistent gender inequalities give rise to slow development, economic growth and poverty reduction (World Bank, 2001). Strategies in developing sustainable ICT-based enterprises on the basis of self-help groups, fostered by macro policies and institutional support from governmental agencies, intermediaries and NGOs. The SL approaches typify a shift in development practice from needs-based, resource- centred solutions to a focus on people and their capacity to initiate and sustain positive change (Carney, 1999). This framework has been widely adapted in various contexts ranging from natural resource management to urban poverty reduction. In this particular context, we wish to adapt it for application to ICT-based enterprises for women. This can be seen as involving two as pects: first the clear incorporation of a gender perspective on livelihoods ; second, incorporation of technology. Each of these will be dealt with in turn. Gender is significant in influencing livelihood outcomes significantly. For example, gender differences in distribution and access to assets – such as credit or technology – are crucial in ownership and management of sustainable enterprises. The UNDP's Social Development and Poverty Eradication Division (SEPED) have integrated technology into its Sustainable Livelihood framework (see UNDP, 1997) as gendered access to and use of technologies is crucial for livelihood outcomes.

In the transition towards a knowledge economy, stakeholders are increasingly required to understand existing and new knowledge systems, for the exchange of information and its transfer into action. A clear understanding of the effect and integration of new ICTs into established knowledge networks (where people are both information providers and seekers) and the contextual issues around the exchange of information, can provide guidelines for new ICTs in rural areas. Poor access to, or inadequate information and knowledge, can significantly hinder poor people from achieving sustainable livelihoods.

Mobile Computing and Access to Information

Nowadays, everything goes mobile. 2013 smartphone is more powerful than a 1990 company's server. There are a lot of opportunities around this movement ranging from manufactures to consumers, including business owners and telecom companies. Everyone can have his niche and generates income. To low level citizens, businesses are doubling efforts to be accessing in very single and possible way through mobile apps and even very basic mobile phones. On the issue of mobile phones, there is a need to develop services that are nuanced towards the real needs of the rural people and incorporate them into the mobile technology. Agricultural extension extend from a process of technology transfer (research institution to farmer) to a process of facilitating a wide range of communication, information, and advocacy services (demand-driven, pluralistic and decentralized extension). In the midst of this change, extensionists are grappling with how best to harness information and communication technologies (ICTs) to improve rural livelihoods. The SL approach shows the variety of activities that people carry out, often in combination, to make a living. It is important in the case of the poor, who often rely on a number of different types of economic activities for their livelihoods, and where it is not any activity but their combined effect for the household economy that matters (Chambers, 1995). Sectoral approaches which tend to pre-determine which area of economic activity people should focus on, for example forestry or fisheries, might lead to only one aspect of people's livelihoods being addressed, and not necessarily the aspect that is most relevant to the poor. SL approach is that it facilitates an understanding of the underlying causes of poverty by focusing on the variety of factors, at different levels, that directly or indirectly determine or constrain poor people's access to resources/assets of different kinds, and thus their livelihoods. Such constraints might spring from formal and informal institutional and social factors at the local level, or they may be the outcome of overriding policies, economic processes, and legislative frameworks at the macro level. A 'micro–macro' perspective is thus built into the approach and is more likely to lead to more strategic interventions.

By focusing on the manner in which people develop their livelihood strategies (coping and adapting strategies) to achieve certain outcomes in response to a particular 'vulnerability context', the SL approach makes it possible to see how even the 'poorest of the poor' are active decision-makers, not passive victims, in shaping their own livelihoods. This is important for designing support activities that build on the strengths of the poor. Also, it allows for a more dynamic perspective on livelihoods, since people's strengths may change over time as their strategies change in response to either personal or external circumstances.

The SL approach facilitates an understanding of the linkages between people's livelihood strategies, their asset status, and their way of using available natural resources, and is therefore a useful approach for understanding both the problem and the scope for promoting sustainable development at the local level. The concept of livelihood offers a more ppropriate basis for evaluating the socio-economic impact of projects or programmes which have poverty alleviation as at least one of their overall objectives, since it provides a more realistic framework for assessing the direct and indirect effects on people's living conditions than, for example, one-dimensional productivity or income criteria.

However, as ICTs continue to diffuse and as greater attempts are made to apply them to current poverty-focused agenda goals there are increasing opportunities for livelihoods frameworks and tools to make a contribution to understanding ICTs and development. This is true of ICT-based enterprise for women though, as seen, some situation-specific customisation of livelihoods tools has to be made. Such customisation is likely to progress further once a body of research is built that does apply livelihoods ideas to ICT-based enterprise.

References

1. DFID 2001. *Sustainable Livelihood Guidance Sheets*, DFID.
2. Carney, D. 1999. *Sustainable Livelihood Approaches Compared*. DFID: London.

3. UNDP 1997. *Sustainable Livelihood Approaches in Operation: A Gender Perspective*. Paper Prepared by Perpetua Katepa-Kalala at the International Associates for Development for the Meeting of the International Working Group on Sustainable Livelihoods. New York.
4. Chambers, R. 1995. *Poverty and livelihoods: Whose reality counts?* Sussex: Institute of Development Studies.

Chapter 8

Statistical Tools for Livelihood Analysis

M. Vijayabhama

Assistant Professor (Statistics),
Department of Forest Resource Management,
Forest College and Research Institute,
Mettupalayam – 641 301, Tamil Nadu

Descriptive statistics is the discipline of quantitatively describing the main features of a collection of data. Descriptive statistics are distinguished from inferential statistics (or inductive statistics), in that descriptive statistics aim to summarize a sample, rather than use the data to learn about the population that the sample of data is thought to represent. This generally means that descriptive statistics, unlike inferential statistics, are not developed on the basis of probability theory. Even when a data analysis draws its main conclusions using inferential statistics, descriptive statistics are generally also presented. Descriptive statistics is set of brief descriptive coefficients that summarize a given data set that represents either the entire population or a sample. The measures that describe the data set are measures of central tendency and measures of variability or dispersion. Measures of central tendency include the mean, median and mode, while measures of variability include the standard deviation (or variance), the minimum and maximum variables, kurtosis and skewness.

In statistics, a frequency distribution is an arrangement of the values that one or more variables take in a sample. Each entry in the table contains the frequency or count of the occurrences of values within a particular group or interval, and in this way, the table summarizes the distribution of values in the sample.

Univariate Frequency Table

Rank	*Degree of Agreement*	*Number*
1	Strongly agree	20
2	Agree somewhat	30
3	Not sure	20
4	Disagree somewhat	15
5	Strongly disagree	15

A different tabulation scheme aggregates values into bins such that each bin encompasses a range of values. For example, the heights of the students in a class could be organized into the following frequency table.

Height Range	*Number of Students*	*Cumulative Number*
4.5–5.0 feet	25	25
5.0–5.5 feet	35	60
5.5–6 feet	20	80
6.0–6.5 feet	20	100

A frequency distribution shows us a summarized grouping of data divided into mutually exclusive classes and the number of occurrences in a class. It is a way of showing unorganized data *e.g.* to show results of an election, income of people for a certain region, sales of a product within a certain period, student loan amounts of graduates, etc. Some of the graphs that can be used with frequency distributions are histograms, line graphs, bar charts and pie charts. Frequency distributions are used for both qualitative and quantitative data.

Arithmetic Mean (AM)

The arithmetic mean is the "standard" average, often simply called the "mean".

$$\bar{x} = \frac{1}{n} \cdot \sum_{i=1}^{n} x_i$$

For example, the arithmetic mean of five values: 4, 36, 45, 50, and 75 is

$$\frac{4+36+45+50+75}{5} = \frac{210}{5} = 42$$

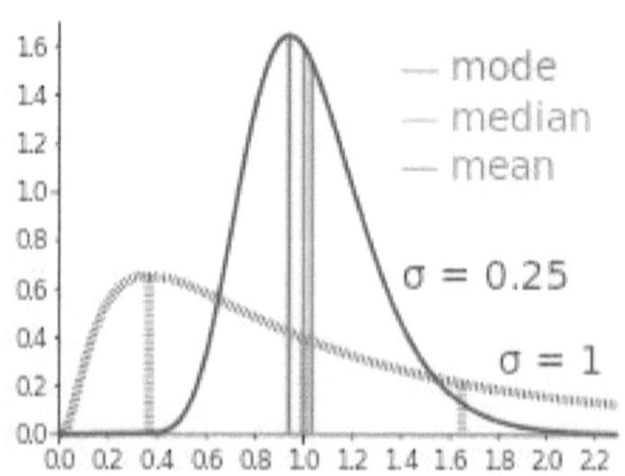

Comparison of the arithmetic mean, median and mode of two skewed (log-normal) distributions.

The mean may often be confused with the median, mode or range. The mean is the arithmetic average of a set of values, or distribution; however, for skewed distributions, the mean is not necessarily the same as the middle value (median), or the most likely (mode). For example, mean income is skewed upwards by a small number of people with very large incomes, so that the majority has an income lower than the mean. By contrast, the median income is the level at which half the population is below and half is above. The mode income is the most likely income, and favors the larger number of people with lower incomes. The median or modes are often more intuitive measures of such data.

Median

Median is defined as middle most items. Since the median is the same as the second quartile,

If **n is odd** then Median (M) = value of $((n + 1)/2)^{th}$ item term.

If **n is even** then Median (M) = value of $[((n)/2)^{th}$ item term + $((n)/2 + 1)^{th}$ item term]/2.

For an Odd Number of Values

As an example, we will calculate the sample median for the following set of observations: 1, 5, 2, 8, 7.

Start by sorting the values: 1, 2, 5, 7, 8.

In this case, the median is 5 since it is the middle observation in the ordered list.

The median is the $((n + 1)/2)^{th}$ item, where n is the number of values. For example, for the list {1, 2, 5, 7, 8}, we have $n = 5$, so the median is the $((5 + 1)/2)^{th}$ item.

median = $(6/2)^{th}$ item

median = 3rd item

median = 5.

For an Even Number of Values

As an example, we will calculate the sample median for the following set of observations: 1, 6, 2, 8, 7, 2.

Start by sorting the values: 1, 2, 2, 6, 7, 8.

In this case, the arithmetic mean of the two middlemost terms is (2 + 6)/2 = 4. Therefore, the median is 4 since it is the arithmetic mean of the middle observations in the ordered list.

We also use this formula MEDIAN = $\{(n + 1)/2\}^{th}$ item. n = number of values

As above example 1, 2, 2, 6, 7, 8 $n = 6$ Median = $\{(6 + 1)/2\}^{th}$ item = 3.5th item. In this case, the median is average of the 3rd number and the next one (the fourth number). The median is

(2 + 6)/2 which is 4.

Mode

The mode is the value that appears most often in a set of data.

The mode of a sample is the element that occurs most often in the collection. For example, the mode of the sample [1, 3, 6, 6, 6, 6, 7, 7, 12, 12, and 17] is 6. Given the list of data [1, 1, 2, 4, 4] the mode is not unique - the dataset may be said to be bimodal, while a set with more than two modes may be described as multimodal.

Mean Median Mode for Raw and Grouped Data

If the case is individual the following are the formulae:-

$$\overline{X} = \frac{\Sigma X}{N} = \frac{x_1 + x_2 + ... + x_n}{n}$$

Median

Case I: When n is odd. In this case the value is the median is,

i.e. $M = \frac{n+1}{2}$ th term

Case II: When n is even $M = \dfrac{\frac{n}{2} + \left(\frac{n}{2} + 1\right)}{2}$

Mode

Most frequently occurred value (or) highest value

If the case is discrete the following are the formulae:-

Mean $= \frac{\Sigma fx}{N}$ Where $N = \Sigma f$

Median

1. Arrange the data in descending or ascending order
2. Find the cumulative frequency

$$M = \frac{n+1}{2}$$

Mode

Form a grouping table if it's necessary

If the Case is Continuous the Following are the Formulae

Mean $= \frac{\Sigma fm}{N}$ Where $N = \Sigma f$, midpoint = m

Median: $M = L + \dfrac{\frac{n}{2} - C}{f} x\, i$

Mode: $M_o = L + \frac{f_s}{f_p + f_s} x\, c$

Quartile

In descriptive statistics, the quartiles of a set of values are the three points that divide the data set into four equal groups, each representing a fourth of the population being sampled. A quartile is a type of quantile. In epidemiology, sociology and finance, the quartiles of a population are the four subpopulations defined by classifying individuals according to whether the value concerned falls into one of the four ranges defined by the three values discussed above. Thus an individual item might be described as being in the upper quartile.

Definitions

First quartile (designated Q_1) = lower quartile = splits lowest 25 per cent of data = 25th percentile.

Second quartile (designated Q_2) = median = cuts data set in half = 50th percentile Third quartile (designated Q_3) = upper quartile = splits highest 25 per cent of data or lowest 75 per cent = 75th percentile.

The difference between the upper and lower quartiles is called the inter quartile range.

Quartile Deviation (Individual)

$$QD = \frac{Q_3 - Q_1}{2}$$

$$Q_3 = \frac{N+1}{4}$$

$Q_1 = 3\{N+1/4\}$

Coefficient of Quartile Deviation = $\frac{Q_3 - Q_1}{Q_3 + Q_1}$

Standard Deviation

$$\sigma = \sqrt{\left[\frac{\sum d^2}{N} - \left(\frac{\sum d}{N}\right)^2\right]}$$

where, d = x - A

If the Case is Discrete the Following are the Formulae

Quartile Deviation

1. Find cumulative frequency
2. Then apply the same formula

Standard Deviation

$$\sigma = \sqrt{\left[\frac{\sum fd^2}{\sum f} - \left(\frac{\sum fd}{\sum f}\right)^2\right]}$$

Where d = x–A

If the Case is Continuous the Following are the Formulae

Quartile Deviation

Q_1 = size of {n/4} th item

$$Q_1 = L + \frac{N/4 - cf}{f} x\, i$$

$$Q_3 = L + \frac{3N/4 - cf}{f} x\, i$$

$$QD = \frac{Q_3 - Q_1}{2}$$

Standard Deviation

$$\sigma = \sqrt{\left[\frac{\sum fX^2}{\sum f} - \left(\frac{\sum fX}{\sum f}\right)^2\right]}$$

The interquartile range is a more useful measure of spread than the range as it describes the middle 50 per cent of the data values.

Deciles

Decile refers to one of ten equal groups which are divided a large group of values or statistics. It is any one of the numbers or values in a series dividing the distribution of the individuals in the series into ten groups of equal frequency. The deciles are the nine values of the variable that divide an ordered data set into ten equal parts. The deciles determine the values for 10 per cent 20 per cent. and 90 per cent of the data. Values 10 per cent, 20 per cent. and 90 per cent D5 coincides with the median. The Decile function computes the specified decile of the specified random variable or data set. The first parameter can be a data set (represented as an Array), a distribution, a random variable, or an algebraic expression involving random variables.

Percentiles

In statistics, a percentile is the value of a variable below which a certain percent of observations fall. For example, the 20th percentile is the value (or score) below which 20 percent of the observations may be found. The term percentile and the related term percentile rank are often used in the reporting of scores from norm-referenced tests. The 25th percentile is also known as the first quartile (Q_1), the 50th percentile as the median or second quartile (Q_2), and the 75th percentile as the third quartile (Q_3).

There is no universally accepted definition of a percentile. Using the 65th percentile as an example, the 65th percentile can be defined as the lowest score that is greater than 65 per cent of the scores. This is the way we defined it above and we will call this "Definition 1". The 65th percentile can also be defined as the smallest score that is greater than or equal to 65 per cent of the scores. This we will call "Definition 2". Unfortunately, these two definitions can lead to dramatically different results, especially when there is relatively little data. Moreover, neither of these definitions is explicit about how to handle rounding. For instance, what score is required to be higher than 65 per cent of the scores when the total number of scores is 50? This is risky because 65 per cent of 50 is 32.5. How do we find the lowest number that is higher than 32.5 of the scores? A third way to compute percentiles is a weighted average of the percentiles computed according to the first two definitions. This third definition handles rounding more gracefully than the other two and has the advantage that it allows the median to be defined conveniently as the 50th percentile.

Standard Deviation and Variance

The variance and the closely-related standard deviation are measures of how spread out a distribution is. In other words, they are measures of variability. The variance and standard deviation are related indicators of the spread of data within a population or sample. The same distinction exists between the population and sample variance. The population variance and standard deviation, denoted σ^2 and σ, are the deviation among individual measurements from the population mean, for the entire population. For the sample variance and standard deviation, s^2 and s, it is how much each individual measurement deviates from the sample mean. As is the case with the mean, the population variance and standard deviation are the expected, or *true*, deviations. The population variance is calculated using the population mean

$$\sigma^2 = \frac{1}{n}\sum_i (x_i - \mu)^2$$

To calculate the sample variance, we first find the errors of all the measurements, that is, the difference between each measurement and the sample mean, $(x_i - \bar{x})$. We then square each value and add them all together, then divide by the number of samples, minus 1. $s^2 = \frac{1}{n-1}\sum_i (x_i - \bar{x})^2$

The sample standard deviation is simply the square-root of the sample variance, $\sqrt{s^2}$.

Correlation

The quantity r, called the linear correlation coefficient, measures the

Strength and the direction of a linear relationship between two variables.

The linear correlation coefficient is sometimes referred to as the Pearson product moment correlation coefficient in honor of its developer Karl Pearson.

Correlation Coefficient

Formula: $$\frac{\sum xy - \frac{\sum x \sum y}{n}}{\sqrt{\sum x^2 - \frac{(\sum x)^2}{n}}\sqrt{\sum y^2 - \frac{(\sum y)^2}{n}}}$$

Correlation Coefficient = r =

Testing the significance of 'r':-

$$t = \frac{r}{\sqrt{1-r^2}}\sqrt{n-2} \sim t_{(n-2)\ df}$$

The value of r is such that $-1 \leq r \leq +1$. The + and – signs are used for positive linear correlations and negative linear correlations, respectively.

Positive Correlation

If x and y have a strong positive linear correlation, r is close to +1. An r value of exactly +1 indicates a perfect positive fit. Positive values indicate a relationship between x and y variables such that as values for x increase, values for y also increase.

Negative Correlation

If x and y have a strong negative linear correlation, r is close to -1. An r value of exactly -1 indicates a perfect negative fit. Negative values indicate a relationship between x and y such that as values for x increase, values for y decrease.

No Correlation

If there is no linear correlation or a weak linear correlation, r is close to 0. A value near zero means that there is a random, nonlinear relationship between the two variables. Note that r is a dimensionless quantity; that is it does not depend on the units employed.

A perfect correlation of ± 1 occurs only when the data points all lie exactly on a straight line. If $r = +1$, the slope of this line is positive. If $r = -1$, the slope of this line is negative. A correlation greater than 0.8 is generally described as strong, whereas a correlation less than 0.5 is generally described as weak. These values can vary based upon the "type" of data being examined. A study utilizing scientific data may require a stronger correlation than a study using social science data.

Coefficient of Determination, r^2 or R^2

The coefficient of determination, r^2, is useful because it gives the proportion of the variance (fluctuation) of one variable that is predictable from the other variable. It is a measure that allows us to determine how certain one can be in making predictions from a certain model/graph. The coefficient of determination is the ratio of the explained variation to the total variation. The coefficient of determination is such that $0 \leq r^2 \leq 1$, and denotes the strength of the linear association between x and y. The coefficient of determination r the percent of the data that is the closest to the line of best fit. For example, if $r = 0.922$, then $r^2 = 0.850$, which means that 85 per cent of the total variation in y can be explained by the linear relationship between x and y (as described by the regression equation). The other 15 per cent of the total variation in y remains unexplained. The coefficient of determination is a measure of how well the regression line represents the data. If the regression line passes exactly through every point on the scatter plot, it would be able to explain all of the variation. The further the line is away from the points, the less it is able to explain.

Correlation coefficient may refer to

- ✰ Pearson product-moment correlation coefficient, also known as r, R, or Pearson's r, a measure of the strength and direction of the linear relationship between two variables that is defined in terms of the (sample) covariance of the variables divided by their (sample) standard deviations. The value of Pearson's correlation coefficient, unlike all other estimates for continuous dependent variables, is influenced by the distribution of the independent variable in the sample.
- ✰ Correlation and dependence, a broad class of statistical relationships between two or more random variables or observed data values.
- ✰ Goodness of fit, which refers to any of several measures that measure how well a statistical model fits observations by summarizing the discrepancy between observed values and the values expected under the model in question.
- ✰ Coefficient of determination, a measure of the proportion of variability in a data set that is accounted for by a statistical model; often called R^2; equal in a single-variable linear regression to the square of Pearson's product-moment correlation coefficient. (The strength of the linear relationship between the values).
- ✰ Intraclass correlation, a descriptive statistic that can be used when quantitative measurements are made on units that are organized into groups; describes how strongly units in the same group resemble each other.
- ✰ Rank correlation, the study of relationships between different rankings on the same set of items.

- Spearman's rank correlation coefficient, which measures how well the relationship between two variables can be described by a monotonic function
- Kendall tau rank correlation coefficient, which measures the portion of ranks that match between two data sets.

Regression Analysis

Regression analysis gives information on the relationship between a response (dependent) variable and one or more (predictor) independent variables to the extent that information is contained in the data. The goal of regression analysis is to express the response variable as a function of the predictor variables. The duality of fit and the accuracy of conclusion depend on the data used. Hence non-representative or improperly compiled data result in poor fits and conclusions. Thus, for effective use of regression analysis one must

1. Investigate the data collection process,
2. Discover any limitations in data collected, and
3. Restrict conclusions accordingly.

Once a regression analysis relationship is obtained, it can be used to predict values of the response variable, identify variables that most affect the response, or verify hypothesized causal models of the response. The value of each predictor variable can be assessed through statistical tests on the estimated coefficients (multipliers) of the predictor variables.

An example of a regression model is the linear regression model which is a linear relationship between response variable, y and the predictor variable, $x_i, i = 1,2..., n$ of the form

$$y = \beta_0 + \beta_1 x_1 + \beta_2 x_2 + ... + \beta_n x_n + \varepsilon$$

where,

$\beta_0, \beta_1 .... \beta_n$ are regression coefficients (unknown model parameters), and ε is the error due to variability in the observed responses.

Regression Definition

A regression is a statistical analysis assessing the association between two variables. It is used to find the relationship between two variables.

Regression Formula

Simple Linear Regression Equations

Formula

The simple linear regression equation of y on x is given by Y = a + bX

The 1st regression coefficient is b_{yx} (or) b

$$\frac{SP(xy)}{SS(x)} = \frac{\sum xy - \frac{\sum x \sum y}{n}}{\sqrt{\sum x^2 - \frac{(\sum x)^2}{n}}}$$

And the intercept, $a = \bar{y} - b\bar{x}$

The simple linear regression equation of x on y is given by X = a + bY

The 2nd regression coefficient is b_{xy} (or) b

b_{xy} = SP(xy) =

$$\frac{SP(xy)}{SS(y)} = \frac{\sum xy - \frac{\sum x \sum y}{n}}{\sqrt{\sum y^2 - \frac{(\sum y)^2}{n}}}$$

And the intercept, $a = \bar{x} - b\bar{y}$

where ,

x and y are the variables.

b = the slope of the regression line

a = the intercept point of the regression line and the y axis.

N = Number of values or elements

X = First Score

Y = Second Score

ΣXY = Sum of the product of first and Second Scores

ΣX = Sum of First Scores

ΣY = Sum of Second Scores

ΣX^2 = Sum of square First Scores

Uses of Regression Analysis

Three uses for regression analysis are for

1. Prediction
2. Model specification and
3. Parameter estimation.

Regression analysis equations are designed only to make predictions. Good predictions will not be possible if the model is not correctly specified and accuracy of the parameter not ensured. However, accurate prediction and model specification require that all relevant variables be accounted in the data and the prediction equation is defined in the correct functional form for all predictor variables.

Parameter estimation is the most difficult to perform because not only in the model required to be correctly specified, the prediction must also be accurate and the data should allow for good estimation. For example, multicolinearity creates a problem and requires that some estimators may not be used. Thus, limitations of data and inability to measure all predictor variables relevant in a study restrict the use of prediction equations.

Multiple Regression Analysis

How to Find Relationship between Variables and Multiple Regression

- ☆ General Purpose
- ☆ Computational Approach
  - Least Squares
  - The Regression Equation
  - Unique Prediction and Partial Correlation
  - Predicted and Residual Scores
  - Residual Variance and R-square
  - Interpreting the Correlation Coefficient R
- ☆ Assumptions, Limitations, and Practical Considerations
  - Assumption of Linearity
  - Normality Assumption
  - Limitations
  - Choice of the number of variables
  - Multicollinearity and matrix ill-conditioning
  - Fitting centered polynomial models
  - The importance of residual analysis

General Purpose

The general purpose of multiple regression (the term was first used by Pearson, 1908) is to learn more about the relationship between several independent or predictor variables and a dependent or criterion variable. For example, a real estate agent might record for each listing the size of the house (in square feet), the number of bedrooms, the average income in the respective neighborhood according to census data, and a subjective rating of appeal of the house. Once this information has been compiled for various houses it would be interesting to see whether and how these measures relate to the price for which a house is sold. For example, we might learn that the number of bedrooms is a better predictor of the price for which a house sells in a particular neighborhood than how "pretty" the house is (subjective rating). We may also detect "outliers," that is, houses that should really sell for more, given their location and characteristics.

Personnel professionals customarily use multiple regression procedures to determine equitable compensation. We can determine a number of factors or dimensions such as "amount of responsibility" (*Resp*) or "number of people to supervise" (*No_Super*) that we believe to contribute to the value of a job. The personnel analyst then usually conducts a salary survey among comparable companies in the market, recording the salaries and respective characteristics (*i.e.*, values on dimensions) for different positions. This information can be used in a multiple regression analysis to build a regression equation of the form:

Salary = .5*Resp + .8*No_Super

Once this so-called regression line has been determined, the analyst can now easily construct a graph of the expected (predicted) salaries and the actual salaries of job incumbents in his or her company. Thus, the analyst is able to determine which position is underpaid (below the regression line) or overpaid (above the regression line), or paid equitably.

In the social and natural sciences multiple regression procedures are very widely used in research. In general, multiple regression allows the researcher to ask (and hopefully answer) the general question "what is the best predictor of." For example, educational researchers might want to learn what are the best predictors of success in high-school. Psychologists may want to determine which personality variable best predicts social adjustment. Sociologists may want to find out which of the multiple social indicators best predict whether or not a new immigrant group will adapt and be absorbed into society.

See also Exploratory Data Analysis and Data Mining Techniques, the General Stepwise Regression topic, and the General Linear Models topic.

Computational Approach

The general computational problem that needs to be solved in multiple regression analysis is to fit a straight line to a number of points.

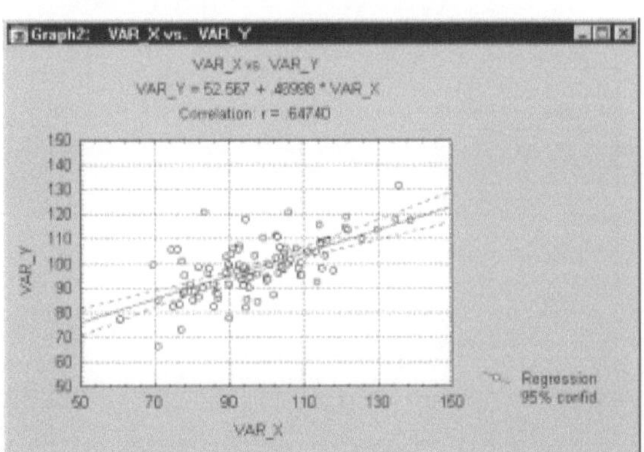

In the simplest case - one dependent and one independent variable - you can visualize this in a scatter plot.

- ☆ Least Squares
- ☆ The Regression Equation
- ☆ Unique Prediction and Partial Correlation
- ☆ Predicted and Residual Scores
- ☆ Residual Variance and R-square
- ☆ Interpreting the Correlation Coefficient R

Least Squares

In the scatter plot, we have an independent or *X* variable, and a dependent or *Y* variable. These variables may, for example, represent IQ (intelligence as measured by a test) and school achievement (grade point average; GPA), respectively. Each point in the plot represents one student, that is, the respective student's IQ and GPA. The goal of linear regression procedures is to fit a line through the points. Specifically, the program will compute a line so that the squared deviations of the observed points from that line are minimized. Thus, this general procedure is sometimes also referred to as least squares estimation.

The Regression Equation

A line in a two dimensional or two-variable space is defined by the equation Y=a+b*X; in full text: the Y variable can be expressed in terms of a constant (a) and a slope (b) times the X variable. The constant is also referred to as the intercept, and the slope as the regression coefficient or B coefficient. For example, GPA may best be predicted as1+.02*IQ. Thus, knowing that a student has an IQ of 130 would lead us to predict that her GPA would be 3.6 (since, 1+.02*130=3.6).

For example, the animation below shows a two dimensional regression equation plotted with three different confidence intervals (90 per cent, 95 per cent and 99 per cent).

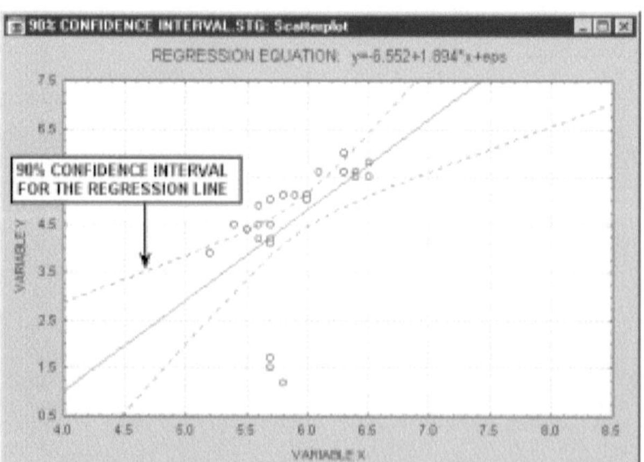

In the multivariate case, when there is more than one independent variable, the regression line cannot be visualized in the two dimensional space, but can be computed just as easily. For example, if in addition to IQ we had additional predictors of achievement (*e.g.*, Motivation, Self- discipline) we could construct a linear equation containing all those variables. In general then, multiple regression procedures will estimate a linear equation of the form:

$$Y = a + b_1{*}X_1 + b_2{*}X_2 +. + b_p{*}X_p$$

Unique Prediction and Partial Correlation

Note that in this equation, the regression coefficients (or *B* coefficients) represent the independent contributions of each independent variable to the prediction of the dependent variable. Another way to express this fact is to say that, for example, variable X_1 is correlated with the Y variable, after controlling for all other independent variables. This type of correlation is also referred to as a partial correlation (this term was first used by Yule, 1907). Perhaps the following example will clarify this issue. We would probably find a significant negative correlation between hair length and height in the population (*i.e.*, short people have longer hair). At first this may seem odd; however, if we were to add the variable Gender into the multiple regression equation, this correlation would probably disappear. This is because women, on the average, have longer hair than men; they also are shorter on the average than men. Thus, after we remove this gender difference by entering Gender into the equation, the relationship between hair length and height disappears because hair length does not make any unique contribution to the prediction of height, above and beyond what it shares in the prediction with variable Gender. Put another way, after controlling for the variable Gender, the partial correlation between hair length and height is zero.

Predicted and Residual Scores

The regression line expresses the best prediction of the dependent variable (Y), given the independent variables (X). However, nature is rarely (if ever) perfectly predictable, and usually there is substantial variation of the observed points around the fitted regression line (as in the scatter plot shown earlier). The deviation of a particular point from the regression line (its predicted value) is called the residual value.

Residual Variance and R-square

R-square, also known as the Coefficient of determination is a commonly used statistic to evaluate model fit. R-square 1 minus the ratio of residual variability. When the variability of the residual values around the regression line relative to the overall variability is small, the predictions from the regression equation are good. For example, if there is no relationship between the X and Y variables, then the ratio of the residual variability of the Y variable to the original variance is equal to 1.0. Then R-square would be 0. If X and Y are perfectly related then there is no residual variance and the ratio of variance would be 0.0, making R-square = 1. In most cases, the ratio and R-square will fall somewhere between these extremes, that is, between 0.0 and 1.0. This ratio value is immediately interpretable in the following manner. If we have an

R-square of 0.4 then we know that the variability of the Y values around the regression line is 1-0.4 times the original variance; in other words we have explained 40 per cent of the original variability, and are left with 60 per cent residual variability. Ideally, we would like to explain most if not all of the original variability. The R-square value is an indicator of how well the model fits the data (*e.g.*, an R-square close to 1.0 indicates that we have accounted for almost all of the variability with the variables specified in the model).

Interpreting the Correlation Coefficient R

Customarily, the degree to which two or more predictors (independent or X variables) are related to the dependent (Y) variable is expressed in the correlation coefficient R, which is the square root of R-square. In multiple regression can assume values between 0 and 1. To interpret the direction of the relationship between variables, look at the signs (plus or minus) of the regression or B coefficients. If a B coefficient is positive, then the relationship of this variable with the dependent variable is positive (*e.g.*, the greater the IQ the better the grade point average); if the B coefficient is negative then the relationship is negative (*e.g.*, the lower the class size the better the average test scores). Of course, if the B coefficient is equal to 0 then there is no relationship between the variables.

Assumptions, Limitations, Practical Considerations

- ✰ Assumption of Linearity
- ✰ Normality Assumption
- ✰ Limitations
- ✰ Choice of the number of variables
- ✰ Multicollinearity and matrix ill-conditioning
- ✰ The importance of residual analysis

Assumption of Linearity

First of all, as is evident in the name multiple linear regression, it is assumed that the relationship between variables is linear. In practice this assumption can virtually never be confirmed; fortunately, multiple regression procedures are not greatly affected by minor deviations from this assumption. However, as a rule it is prudent to always look at bivariate scatter plot of the variables of interest. If curvature in the relationships is evident, we may consider either transforming the variables, or explicitly allowing for nonlinear components.

Normality Assumption

It is assumed in multiple regression that the residuals (predicted minus observed values) are distributed normally (*i.e.*, follow the normal distribution). Again, even though most tests (specifically the F-test) are quite robust with regard to violations of this assumption, it is always a good idea, before drawing final conclusions, to review the distributions of the major variables of interest. We can produce histograms for the residuals as well as normal probability plots, in order to inspect the distribution of the residual values.

Limitations

The major conceptual limitation of all regression techniques is that we can only ascertain relationships, but never be sure about underlying causal mechanism. For example, we would find a strong positive relationship (correlation) between the damage that a fire does and the number of firemen involved in fighting the blaze. Do we conclude that the firemen cause the damage? Of course, the most likely explanation of this correlation is that the size of the fire (an external variable that we forgot to include in our study) caused the damage as well as the involvement of a certain number of firemen (*i.e.*, the bigger the fire, the more firemen are called to fight the blaze). Even though this example is fairly obvious, in real correlation research, alternative causal explanations are often not considered.

Choice of the Number of Variables

Multiple Regression is a seductive technique: "plug in" as many predictor variables as we can think of and usually at least a few of them will come out significant. This is because we are capitalizing on chance when simply including as many variables as we can think of as predictors of some other variable of interest. This problem is compounded when, in addition, the number of observations is relatively low. Intuitively, it is clear that we can hardly draw conclusions from an analysis of 100 questionnaire items based on 10 respondents. Most authors recommend that we should have at least 10 to 20 times as many observations (cases, respondents) as we have variables; otherwise the estimates of the regression line are probably very unstable and unlikely to replicate if we were to conduct the study again.

Multicollinearity and Matrix Ill-Conditioning

This is a common problem in many correlation analyses. Imagine that we have two predictors (X variables) of a person's height: (1) weight in pounds and (2) weight in ounces. Obviously, our two predictors are completely redundant; weight is one and the same variable, regardless of whether it is measured in pounds or ounces. Trying to decide which one of the two measures is better predictor of height would be rather silly; however, this is exactly what we would try to do if we were to perform a multiple regression analysis with height as the dependent (Y) variable and the two measures of weight as the independent (X) variables. When there are very many variables involved, it is often not immediately apparent that this problem exists, and it may only manifest itself after several variables have already been entered into the regression equation. Nevertheless, when this problem occurs it means that at least one of the predictor variables is (practically) completely redundant with other predictors. There are many statistical indicators of this type of redundancy (tolerances, semi-partial R, etc., as well as some remedies (*e.g.*, Ridge regression).

Fitting Centered Polynomial Models

The fitting of higher-order polynomials of an independent variable with a mean not equal to zero can create difficult multicollinearity problems. Specifically, the polynomials will be highly correlated due to the mean of the primary independent variable. With large numbers (*e.g.*, Julian dates), this problem is very serious, and if proper protections are not put in place, can cause wrong results. The solution is to "center" the independent variable (sometimes, this procedures is referred to as "centered polynomials"), *i.e.*, to subtract the mean, and then to compute the polynomials. See, for example, the classic text by Neter, Wasserman, & Kutner (1985, Chapter 9), for a detailed discussion of this issue (and analyses with polynomial models in general).

The Importance of Residual Analysis

Even though most assumptions of multiple regression cannot be tested explicitly, gross violations can be detected and should be dealt with appropriately. In particular outliers (*i.e.*, extreme cases) can seriously bias the results by "pulling" or "pushing" the regression line in a particular direction (see the animation below), thereby leading to biased regression coefficients. Often, excluding just a single extreme case can yield a completely different set of results.

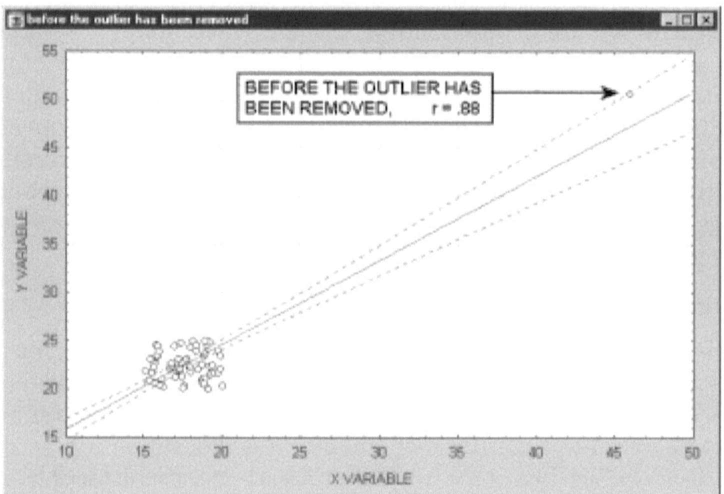

Step-wise Method

Step-up Regression

At any stage, rather than entering all the variables as a block, step-up regression enters the variables one at a time, the order of entry determined by the variable that causes the greatest R^2 increase, given the variables already entered into the model. To do a step-up regression using SPSS/WIN, enter all the variables in the first block and select "Method" as "Forward."

The results of the step-up regression can be better understood if the correlation coefficients are recomputed between life satisfaction seven years after college and all the predictor variables, using the "List-wise" option for missing data

Step-down Regression

By starting with a full model and eliminating variables that do not significantly enter the regression equation, a partial model may be found. This can be accomplished in SPSS/WIN by selecting a "Method" of "Backward" in the linear regression procedure. As can be seen below, the results of this analysis differ greatly from the use of the Forward Method.

Step-wise Regression

In this section, we learn about the stepwise regression procedure. While we will soon learn the finer details, the general idea behind the stepwise regression procedure is that we build our regression model from a set of candidate predictor variables by entering and removing predictors in a stepwise manner into our model until there is no justifiable reason to enter or remove any more.

Our hope is, of course, that we end up with a reasonable and useful regression model. There is one sure way of ending up with a model that is certain to be underspecified and that's if the set of candidate predictor variables doesn't include *all* of the variables that actually predict the response. This leads us to a fundamental rule of the stepwise regression procedure the list of candidate predictor variables must include *all* of the variables that actually predict the response. Otherwise, we are sure to end up with a regression model that is underspecified and therefore misleading.

Factor Analysis

Factor analysis a statistical method used to describe variability describe variability among observed, correlated variables in terms of a potentially lower number of unobserved variables called factors. In other words, it is possible, for example, that variations in three or four observed variables mainly reflect the variations in fewer unobserved variables. Factor analysis searches for such joint variations in response to unobserved latent variables. The observed variables are modeled as linear combinations of the potential factors, plus "error" terms. The information gained about the interdependencies between observed variables can be used later to reduce the set of variables in a dataset. Computationally this technique is equivalent to low rank approximation of the matrix of observed variables. Factor analysis originated in psychometrics, and is used in behavioral sciences, social sciences, marketing, product management, operations research, and other applied sciences that deal with large quantities of data.

Factor analysis is related to principal component analysis (PCA), but the two are not identical. Latent variable models, including factor analysis, use regression modeling techniques to test hypotheses producing error terms, while PCA is a descriptive statistical technique. There has been significant controversy in the field over the equivalence or otherwise of the two techniques (see exploratory factor analysis versus principal components analysis).

Type of Factor Analysis

- ☆ **Exploratory factor analysis (EFA)** is used to uncover the underlying structure of a relatively large set of variables. The researcher's a priori assumption is that any indicator may be associated with any factor. This is the most common form of factor analysis. There is no prior theory and one uses factor loadings to intuit the factor structure of the data.

- ☆ **Confirmatory factor analysis (CFA)** seeks to determine if the number of factors and the loadings of measured (indicator) variables on them conform to what is expected on the basis of pre-established theory. Indicator variables are selected on the basis of prior theory and factor analysis is used to see if they load as predicted on the expected number of factors. The researcher's a priori assumption is that each factor (the number and labels of which may be specified a priori) is associated with a specified subset of indicator variables. A minimum requirement of confirmatory factor analysis is that one hypothesizes beforehand the number of factors in the model, but usually also the researcher will posit expectations about which variables will load on which factors. The researcher seeks to determine, for instance, if measures created to represent a latent variable really belong together.

Types of Factoring

- ☆ **Principal component analysis (PCA)**: PCA seeks a linear combination of variables such that the maximum variance is extracted from the variables. It then removes this variance and seeks a second linear combination which explains the maximum proportion of the remaining variance, and so on. This is called the principal axis method and results in orthogonal (uncorrelated) factors.
- ☆ **Canonical factor analysis**, also called Rao's canonical factoring, is a different method of computing the same model as PCA, which uses the principal axis method. Canonical factor analysis seeks factors which have the highest canonical correlation with the observed variables. Canonical factor analysis is unaffected by arbitrary rescaling of the data.
- ☆ **Common factor analysis**, also called principal factor analysis (PFA) or principal axis factoring (PAF), seeks the least number of factors which can account for the common variance (correlation) of a set of variables.
- ☆ **Image factoring:** based on the correlation matrix of predicted variables rather than actual variables, where each variable is predicted from the others using multiple regression.
- ☆ **Alpha factoring:** based on maximizing the reliability of factors, assuming variables are randomly sampled from a universe of variables. All other methods assume cases to be sampled and variables fixed.
- ☆ **Factor regression model:** a combinatorial model of factor model and regression model; or alternatively, it can be viewed as the hybrid factor model, whose factors are partially known.

Terminology

Factor loadings: The factor loadings, also called component loadings in PCA, are the correlation coefficients between the variables (rows) and factors (columns).

Analogous to Pearson's r, the squared factor loading is the percent of variance in that indicator variable explained by the factor. To get the percent of variance in all the variables accounted for by each factor, add the sum of the squared factor loadings for that factor (column) and divide by the number of variables. (Note the number of variables equals the sum of their variances as the variance of a standardized variable is 1). This is the same as dividing the factor's eigen value by the number of variables.

Interpreting factor loadings: By one rule of thumb in confirmatory factor analysis, loadings should be 7 or higher to confirm that independent variables identified a priori are represented by a particular factor, on the rationale that the 7 level corresponds to about half of the variance in the indicator being explained by the factor. However, the 7 standard is a high one and real-life data may well not meet this criterion, which is why some researchers, particularly for exploratory purposes, will use a lower level such as.4 for the central factor and.25 for other factors call loadings above 6 "high" and those below 4 "low". In any event, factor loadings must be interpreted in the light of theory, not by arbitrary cutoff levels.

In oblique rotation, one gets both a pattern matrix and a structure matrix. The structure matrix is simply the factor loading matrix as in orthogonal rotation, representing the variance in a measured variable explained by a factor on both a unique and common contributions basis. The pattern matrix, in contrast, contains coefficients which just represent unique contributions. The more factors, the lower the pattern coefficients as a rule since there will be more common contributions to variance explained. For oblique rotation, the researcher looks at both the structure and pattern coefficients when attributing a label to a factor.

Communality: The sum of the squared factor loadings for all factors for a given variable (row) is the variance in that variable accounted for by all the factors, and this is called the communality. The communality measures the percent of variance in a given variable explained by all the factors jointly and may be interpreted as the reliability of the indicator.

Spurious solutions: If the communality exceeds 1.0, there is a spurious solution, which may reflect too small a sample or the researcher has too many or too few factors.

Uniqueness of a variable: That is, uniqueness is the variability of a variable minus its communality.

Eigen values:/Characteristic roots: The eigen value for a given factor measures the variance in all the variables which is accounted for by that factor. The ratio of eigen values is the ratio of explanatory importance of the factors with respect to the variables. If a factor has a low eigen value, then it is contributing little to the explanation of variances in the variables and may be ignored as redundant with more important factors. Eigen values measure the amount of variation in the total sample accounted for by each factor.

Extraction sums of squared loadings: Initial eigen values and eigen values after extraction (listed by SPSS as "Extraction Sums of Squared Loadings") are the same for PCA extraction, but for other extraction methods, eigen values after extraction will be lower than their initial counterparts. SPSS also prints "Rotation Sums of Squared Loadings" and even for PCA, these eigen values will differ from initial and extraction eigen values, though their total will be the same.

Factor scores (also called component scores in PCA): are the scores of each case (row) on each factor (column). To compute the factor score for a given case for a given factor, one takes the case's standardized score on each variable, multiplies by the corresponding factor loading of the variable for the given factor, and sums these products. Computing factor scores allows one to look for factor outliers. Also, factor scores may be used as variables in subsequent modeling.

Criteria for determining the number of factors: Using one or more of the methods below, the researcher determines an appropriate range of solutions to investigate. Methods may not agree. For instance, the Kaiser criterion may suggest five factors and the screen test may suggest two, so the researcher may request 3-, 4-, and 5-factor solutions discuss each in terms of their relation to external data and theory.

Comprehensibility: A purely subjective criterion would be to retain those factors whose meaning is comprehensible to the researcher. This is not recommended

Kaiser criterion: The Kaiser rule is to drop all components with eigen values under 1.0 – this being the eigen value equal to the information accounted for by an average single item. The Kaiser criterion is the default in SPSS and most statistical software but is not recommended when used as the sole cut-off criterion for estimating the number of factors as it tends to over extract factors.

Variance explained criteria: Some researchers simply use the rule of keeping enough factors to account for 90 per cent (sometimes 80 per cent) of the variation. Where the researcher's goal emphasizes parsimony (explaining variance with as few factors as possible), the criterion could be as low as 50 per cent.

Scree plot: The Cattell scree test plots the components as the X axis and the corresponding eigen values as the Y-axis. As one moves to the right, toward later components, the eigenvalues drop. When the drop ceases and the curve makes an elbow toward less steep decline, Cattell's scree test says to drop all further components after the one starting the elbow. This rule is sometimes criticized for being amenable to researcher-controlled "fudging". That is, as picking the "elbow" can be subjective because the curve has multiple elbows or is a smooth curve, the researcher may be tempted to set the cut-off at the number of factors desired by his or her research agenda.

Horn's Parallel Analysis (PA): A Monte-Carlo based simulation method that compares the observed eigen values with those obtained from uncorrelated normal variables. A factor or component is retained if the associated eigen value is bigger than the 95th of the distribution of eigen values derived from the random data. PA is one of the most recommendable rules for determining the number of components to retain, but only few programmes include this option.

Before dropping a factor below one's cut-off, however, the researcher should check its correlation with the dependent variable. A very small factor can have a large correlation with the dependent variable, in which case it should not be dropped.

Rotation methods: The unrotated output maximizes the variance accounted for by the first and subsequent factors, and forcing the factors to be orthogonal. This data-compression comes at the cost of having most items load on the early factors, and usually, of having many items load substantially on more than one factor. Rotation serves to make the output more understandable, by seeking so-called "Simple Structure": A pattern of loadings where items load most strongly on one factor, and much more weakly on the other factors. Rotations can be orthogonal or oblique (allowing the factors to correlate).

Varimax rotation: It is an orthogonal rotation of the factor axes to maximize the variance of the squared loadings of a factor (column) on all the variables (rows) in a factor matrix, which has the effect of differentiating the original variables by extracted factor. Each factor will tend to have either large or small loadings of any particular variable. A varimax solution yields results which make it as easy as possible to identify each variable with a single factor. This is the most common rotation option. However, the orthogonality (*i.e.*, independence) of factors is often an unrealistic assumption. Oblique rotations are inclusive of orthogonal rotation, and for that reason, oblique rotations are a preferred method.

Quartimax rotation: Iis an orthogonal alternative which minimizes the number of factors needed to explain each variable. This type of rotation often generates a general factor on which most variables are loaded to a high or medium degree. Such a factor structure is usually not helpful to the research purpose.

Equimax rotation: It is a compromise between Varimax and Quartimax criteria.

Direct oblimin rotation: It is the standard method when one wishes a non-orthogonal (oblique) solution – that is, one in which the factors are allowed to be correlated. This will result in higher eigen values but diminished interpretability of the factors.

Promax rotation: It is an alternative non-orthogonal (oblique) rotation method which is computationally faster than the direct oblimin method and therefore is sometimes used for very large datasets.

Plates

Plate 1. The Rich Agro-biodiversity of The Nilgiris.

Plate 2. Discussion with Tribal Women of Kothars.

Plate 3. Seed Storage by an Irular Woman.

Plate 4. Soil and Water Conservation of the Tribal Women.

Plate 5. Natural Farming with Organic Inputs.

Plate 6. Land Preparation.

Plate 7. Women in Postharvest Storage of the Produce.

References

Achanta, L.D. 1982. Role of Rural Women in Agricultural Development. *Kurukshetra*, 31(2): 15 and 19.

Anilkumar. 1995. Agro-biodiversity Conservation Practices in the Tribal Areas of Wayanad District, Kerala, India. Concept Publishing House, New Delhi.

*Anonymous. 1996. Report of the International Technical Conference on Plant Genetic Resources, Leipzig, Germany, 17-23 June, Food and Agricultural Organisation of the United Nations, Rome.

Anonymous. 2002. Environment Information System, Centre for Symbiosis of Technology, Bangalore, India.

Anonymous. 2003. Environment and Management, Centre for Symbiosis of Technology, Bangalore, India.

*Anonymous, 2005. World Conference on Women. Proceedings of the Beijing Declaration, Beijing.

Arunachalam, R. 2002. Empowerment of Women and Ecological Development. Serials Publications, New Delhi, pp. 8-9.

Balasubramanian, P. 1992. Indigenous Knowledge Use in Dry Lands: An Exploratory Study. Unpub. M.Sc. (Ag.) Thesis, TNAU, Coimbatore.

Bhuvaneshwari, Sri Vara Buddhi. 2005. Role of Tribal Women in the Conservation of Agro-biodiversity: A Multidimensional Analysis. Unpub. Ph.D. Thesis, TNAU, Coimbatore.

Bhuyan, B and S.K. Tripathy. 1988. Role of Women in Agriculture- A study in Ganjam District of Orissa. *Journal of Rural Development, 7(4): 445-452.*

Brahmi, 2004. The Biological Diversity Act of India and Agro-biodiversity Management. *Current Science. 86(5): 659-664.*

*Bunning, S. and C. Hill. 1996. Farmers Rights in the Conservation and Use of Plant Genetic Resources; A Gender Perspective. Paper presented at the second extraordinary session of the FAO commission on Genetic Resources for Food and Agriculture, Rome.

Chakrabarthy, T. 1997. Economic Participation of Tribal Women of Tripura- A Case Study. Tripura State Cultural Research Institute and Museum, Govt of Tripura, pp 1-213.

Chelala, Cesar. 2002. The Impact of Environment on Children's Health. *Women and Environment, June.*

Deshpande, 1988. Tribals of Tripura. *Indian Farming, 38(8):33.*

Eldredge, N and Salthe, S.N. 1984. Hierarchy and Evolution: Oxford Surveys in Evolutionary Biology. Oxford Press, New Delhi.

FAO. 1999. Report on Sustainable Livelihood, Rome.

FAO. 2006. Annual Report of the Food and Agriculture Organisation, Rome.

Ganadeepa, A. 1991. Techno-cultural Profile of Rice Farmers. Unpub. M.Sc. (Ag.) Thesis, TNAU, Coimbatore.

Ganesan, R. 1989. Participation Pattern of Officials, Farm Leaders and Farmer Beneficiaries in Agricultural Development Schemes – A Critical Analysis, Unpub. Ph.D. Thesis, TNAU, Coimbatore.

*Gari, Joseph, 2001. Agro-biodiversity and Food Security in Sub- Saharan Africa. Report submitted on Strategic Issues for Agricultural Policy in Africa.

Guttman, L. 1944. A Basis for Scaling Quantitative Data: Readings in Attitude Theory and Measurement, John Wiley and Sons. Inc., New York.

IISD. 2005. Annual Report of International Institute for Sustainable Development. USA.

Jamatia, Phani Bhusan. 1999. Participation of Tribal Women in Agriculture and Allied Activities in Tripura State. Unpub. M.Sc. (Ag.) Thesis, TNAU, Coimbatore.

Johnson, M. 1983. The Labyrinth of Community Participation: Experience in Indonesia. *Journal of Community Development, 17(2): 202-207.*

Kaliappan, U.R. 1985. A Study of Status and Role of Farm Women in Coimbatore District of Tamil Nadu. Unpub Ph.D Thesis, Department of Sociology, Bharathiyar University, Coimbatore.

Kokate, R. 1988. Factors Influencing Feeding Gap in Local Cattle of Tribal Scenario. *Journal of Rural Development, 7(6):719-72.*

Kumari, Baby. P. 1998. Farmers and Farm Women in Farm and Home Decision-Making. Unpub. Ph.D. Thesis, TNAU, Coimbatore.

Kumary, Leena. 2008. Sustainable Management of Agro-biodiversity for Food Security in Kerala. Proceedings of National Seminar on Recent Trends in Agro-biodiversity Conservation and Documentation, Mahe, UT of Puducherry.

Lakshmana, Kella. 2000. Documentation and Analysis of Indigenous Farm Practices on Major Crops- An Inventory. Unpub. M.Sc. (Ag.) Thesis, TNAU, Coimbatore

Likert, R. 1949. Technique for the Measurement of Attitudes. *Arch. Psychology, pp.140.*

Madivanane, S. 1990. A Study on Training Needs of Farm Women in Karaikal Region, Pondicherry. Unpub. M.Sc. (Ag.) Thesis, TNAU, Coimbatore.

Mahapatra, S. 1985. Modernization of Tribal Agriculture: Technological and Cultural Restraints. *Economic and Political Weekly, 13(13): 581-585.*

Majumdar, V. 1975. Women in Agriculture. *Indian Farming, 25(3): 6.*

Manjula, L. 1991. Tribal Farm Families- A Critical Study. Unpub. M.Sc. (Ag.) Thesis, TNAU, Coimbatore.

Mansingh, J. Paul. 1993. Construction and Standardization of Socio-Economic Status Scale. Unpub. Ph.D. Thesis, TNAU, Coimbatore.

Margules, C.R. 2002. Representing Biodiversity: Data and Procedures for Identifying Priority Areas for Conservation. John Wiley and Sons, New York.

Marimuthu, P. 1998. Marketing Behaviour of Banana Growers. Unpub. M.Sc. (Ag.) Thesis, TNAU, Coimbatore.

Marimuthu, P. 2001. Indigenous Tribal Wisdom for Rural Development: A Multidimensional Analysis. Unpub. Ph.D. Thesis. TNAU, Coimbatore.

Misra, Suresh and Charanjeen Singh. 1998. Rural Leadership in Haryana : Changing Profile. *Kurukshetra, XLVI(7): pp42-43.*

Mohanty, S.C. 1990. Paroja: Tribes of Orissa. Harijan and Tribal Welfare Department, Government of Orissa, Bhuvaneshwar, pp.221-229.

Nandini, N. 1995. Impact of Soil and Water Conservation Programme Implemented by an NGO and Government Organisation- A Comparative Study. Unpub. M.Sc. (Ag.) Thesis, TNAU, Coimbatore.

Oberoi, R.C. and T.V. Moorthi. 1986. Economics of Tribal Agriculture: A Study of Bharmaur Tehsil in Himachal Pradesh. *Agricultural Situation in India, 41(8): pp 629-634.*

Palmurugan, M. 2002. Prospects of Empowering Farm Women in Irrigated Farming System. Unpub. M.Sc. (Ag.) Thesis, TNAU, Coimbatore.

Pandey, H. 2004. Role of Women in Agricultural Biodiversity Conservation. *Indian Farming, 54(8): 51&52.*

Pareek, Udai and T. Venkateswara Rao. 1974. Hand Book of Psychological and Social Instruments, Samasthi Publications, Baroda.

Parvathi, S. 1995. A Field Experiment on Farm Women's Cognitive Domain Relating to Post Harvest Technology. Unpub Ph.D. Thesis. TNAU, Coimbatore.

Patel, J.B. and B.N. Sandagi. 1995. Work and Time Analysis of Farm Women in Farming System Operations, *Man Power Journal. Vol. XXXI(2): 19-27.*

Patnaik, P.S. 1990. Kandha Tribes of Orissa. Harijan and Tribal Welfare Department. Government of Orissa, Bhuvaneshwar, pp.140-146.

Pratap, D.R. 1998. Andhra Pradesh Jatulu. Tribal Cultural Research and Training Institute, Hyderabad.

Premavathi, R. 1997. Rural Women in Farm and Home Decision Making. Unpub M.Sc (Ag.) Thesis, TNAU, Coimbatore.

Rajulashanthy, T. 1991. Indigenous Resources Management by Farm Women: An Analysis. Unpub M.Sc (Ag.) Thesis, TNAU, Coimbatore.

Ramachandra, T.V. and A.V. Nagarathna, 2003. Biodiversity Characterization and Landscape Level in North-East India using Satellite Remote Sensing and Geographical Information System, *Current Science. Vol.84 (8): pp1141-1143.*

Ramakrishnan, N. 2002. *In situ* and *Ex situ Conservation* of Medicinal Plant Diversity in and around Thiruvanamalai Hills. Paper presented in the National Seminar on Conservation of the Eastern Ghats. Madurai Kamaraj University, Madurai.

Ramakrishnan, N. and A. L. Chidambaram, 2003. A Survey of Ethno-Medicinal plants of Thiruvanamalai Hill. National Seminar on Emerging Trends in Plant Science. Annamalai Nagar, Chidambaram.

Ramanna, K.N. 1999. People's Participation in Planning and Implementation of Intergrated Watershed Development Programme – A Comparative Study of Watersheds of Government *vis-à-vis* NGO. Unpub Ph.D Thesis, TNAU, Coimbatore.

Ramu, S. 1997. Tribal Economy and Plantation Agriculture among the Kurumbas of The Nilgiris Hills. Unpub Ph.D Thesis, Tribal Research Centre, Udagamandalam, Tamil Nadu.

Rani, A. Celine. 2000. Impact of Socio-Economic Status on the Emergence of Rural Women Leadership. *Journal of Extension and Research,* 11(2): pp. 109-122.

Ravishankar, T. and V. Selvam. 1996. Contributions of Tribal Communities in the Conservation of Traditional Cultivars Using Diversity: Enhancing and Maintaining Genetic Resources on- Farm, International Development Research Centre, New Delhi.

Ravishankar,T. 1994. Role of Tribal Communities in the Conservation of Plant Genetic Resources. M. S. Swaminathan Research Foundation, Chennai.

Ray, G.L. 1999. Research Methods in Social Sciences and Extension Education. Naya Prakash publications, Calcutta.

Sailaja, V. 2002. Empowerment of Rural Women through Participation in Co-operative Institutions of Andhra Pradesh – An Analysis. Unpub. M.Sc.(Ag.) Thesis, TNAU, Coimbatore.

Sakunthalai, A. 2004. Identification and Standardization of Critical Indicators to Measure Women Empowerment. Unpub. Ph.D., Thesis, TNAU, Coimbatore.

Santha Govind, 1984. Participation of Farm Women in Farm and Home Activities. Unpub. M.Sc. (Ag.) Thesis. TNAU, Coimbatore.

Santha, E.K. 1999. Political Participation of Women in Panchayat Raj: Haryana, Kerala and Tamil Nadu. Indian Institute of Social Sciences, New Delhi.

Saraswathi, T. 1993. Awareness and Participation of Malayali Tribe in Integrated Tribal Development Programme and its Impact. Unpub. M.Sc. (Ag.) Thesis, TNAU, Coimbatore.

Saravanakumar, M. 2000. Impact of TANWA in the Empowerment of Farm Women. Unpub. M.Sc. (Ag.) Thesis, TNAU, Coimbatore.

Sarojini, M. and U.N. Moorthy. 2002. Empowerment of Women and Ecological Development. Serials Publications, New Delhi.

Seema, B. and R. N. Prasad. 1991. Role Analysis in Decision Making by Farm Women. *Tamil Nadu Journal of Extension Education, 2(4): 344.*

*Sehgal, Ninika. 2004. Role of Women in Environment Conservation, Centre for Symbiosis of Technology, Publications 172, Bangalore.

Selvarani, G. 2000. Training Needs of Tribal Farmers of The Nilgiris. Unpub M.Sc. (Ag.) Thesis, TNAU, Madurai.

Senthamarai, G. 1997. Socio-Personal and Psychological Characteristics of Farm Women. *Journal of Extension Education. Vol(5): 1607-1608*.

Shanthasheela, M. 2002. Performance Analysis of Elected Women Presidents in Village Panchayats. Unpub. Ph.D. Thesis, TNAU, Coimbatore.

Sharma Y.P., 1997. Environmental Education and Role of Mass Media, Paper presented in the National Seminar on Environmental Education on 16-18 October, New Delhi.

Sharma, Mahesh and Rajendra Prasad. 1983. Issues in Tribal Development and the Role of Science and Technology-II. *Changing Villages*, 5(4): 243-256.

Sharma, S.D. 1997. Gender Dimensions in Biodiversity Management in India. Report Submitted to FAO Regional Office for Asia and the Pacific, Bangkok, Thailand.

Singh Naresh and Jonathan Gilman, 2004. Employment and Natural Resource Management: A Livelihood Approach to Poverty Reduction, Conference paper series, United Nations Development Programme (UNDP).

*Singh Naresh, 1996. Community Adaptation and Sustainable Livelihoods: Basic issues and principles, working paper submitted to International Institute of Sustainable Development (IISD), Canada.

Singh, M.P. and Neeraj Sharma. 1993. Hill Women of U.P.: A Survey. *Indian Farming, 38 (3): 8.*

Sripal, K.B. 1983. Tribal Development Programme- A Critical Analysis. Unpub. Ph.D Thesis. TNAU, Coimbatore.

Stapp, William. B. 1982. An Instructional Programme Approach to Environmental Education. Edited by Oxford and IBH publishing Co, New Delhi.

Sujatha, J. 1996. Gender Analysis in Different Farming System. Unpub. Ph.D Thesis. TNAU, Coimbatore.

Supe, S.V. 1969. Factors Related to Different Degree of Rationality in Decision-Making Among Farmers. Unpub. Ph.D. Thesis, IARI, New Delhi.

Swaminathan, M. S. and S. Jana. 1992. Biodiversity Implications for Global Food Security. Macmillan India Ltd, Chennai.

The Convention on Biological Diversity. 1998. Outline for Biological Diversity Act. Ministry of Environment and Forests, Government of India.

Thenamudha, V. 1996. Entrepreneurial Behaviour of Farmers – A Critical Analysis. Unpub. M.Sc. (Ag.) Thesis, TNAU, Coimbatore.

Theodore, R.K. 1999. Farm Diversification for Sustainable Agriculture (FDSA) – An Empirical Analysis. Unpub. Ph.D. Thesis, TNAU, Coimbatore.

Thurstone, L. 1947. Multiple Factor Analysis. University of Chicago Press, Chicago.

Tomar, Archana. 2007. Environmental Education. Kalpaz Publication, New Delhi.

Trivedi, G. 1963. Measurement and Analysis of Socio-economic Status of Rural Families. Unpub. Ph.D. Thesis, IARI, New Delhi.

Trivedi, P.C.2006. Biodiversity: Assessment and Conservation, Agro-bios (India), Jodhpur.

UNESCO. 1990. Source Book in Environmental Education for Secondary School Teachers, Bangkok.

Venkataramaiah, P. 1983. Impact of High Yielding Varieties Programme on the Dissemination of Scientific Agricultural Information among the Farmers of Ganjam District. Unpub. M.Sc. (Ag.) Thesis, OUAT, Bhuvaneshwar.

Warren, D.M. 1991. Indigenous Knowledge in Agricultural Development. *Journal of Extension System, 4(1): 71-82.*

Williams, D. 2005. Data Requirements and Data Sources for Biodiversity: Priority Area Selection (PAS). National Academy Press, Washington.

---

* Original not seen

Appendices

APPENDIX I

Population of Scheduled Tribes in Tamil Nadu by Districts.

Sl.No.	*District*	*Total Population*	*Schedule Tribes*			*Per cent to District Total*
			Rural	*Urban*	*Total*	
1.	Chennai	4343645	0	6728	6728	0.15
2.	Kancheepuram	2877468	18062	8446	26508	0.92
3.	Thiruvallur	2754756	28885	8973	37858	1.37
4.	Cuddalore	2285395	7241	4532	11773	0.52
5.	Villupuram	2960373	61687	2233	63920	2.16
6.	Vellore	3477317	58237	4803	63040	1.81
7.	Thiruvannamalai	2186125	69198	3562	72760	3.33
8.	Salem	3016346	98722	5199	103921	3.45
9.	Namakkal	1493462	50454	962	51416	3.44
10.	Dharmapuri	2856300	57763	1786	59549	2.08
11.	Erode	2581500	15120	2573	17693	0.69
12.	Coimbatore	4271856	19559	9544	29103	0.68
13.	The Nilgiris	762141	19600	8773	28373	3.72
14.	Thiruchirapalli	2418366	14383	4529	18912	0.78
15.	Karur	935686	1075	375	1450	0.15
16.	Perambalur	493646	2768	539	3307	0.67
17.	Ariyalur	695524	7907	622	8529	1.23
18.	Pudukottai	1459601	432	360	792	0.05
19.	Thanjavur	2216138	1302	2339	3641	0.16
20.	Nagapatinam	1488839	1618	1802	3420	0.23
21.	Thiruvarur	1169474	971	1702	2673	0.23
22.	Madurai	2578201	2054	3918	5972	0.23
23.	Theni	1093950	1046	640	1686	0.15
24.	Dindigul	1923014	3512	2972	6484	0.34
25.	Ramanathapuram	1187604	396	682	1078	0.09
26.	Virudhunagar	1751301	953	1404	2357	0.13
27.	Sivagangai	1155356	375	708	1083	0.09
28.	Thirunelveli	2723988	3202	5156	8358	0.31
29.	Thoothukudi	1572273	1060	2434	3494	0.22
30.	Kanyakumari	1676034	3561	1882	5443	0.32
	STATE	62405679	551143	100178	651321	1.04

Source: Directorate of Census Operations, Tamil Nadu 2001 CENSUS.

APPENDIX II

Distribution of Irula Tribal Habitations in the Kothagiri Taluk (2000).

Sl.No.	*Village Name*	*Total Irula Habitation*
1.	Thengumarahada	1
2.	Hallimayar	1
3.	Kallampalayam	1
4.	Kodanad	2
5.	Denad	5
6.	Nandipuram	1
7.	Kadinamala	6
8.	Arakodu	6
9.	Kokkudu	3
10.	Kengarai	5
11.	Jakkanarai	9
12.	Konnavakarai	7
	Total	47

Source: District Adi Dravidar and Tribal Welfare Office, Ootacamund, 2000.

APPENDIX III
Land Use Pattern of The Nilgiris District.

Sl.No.	*Land Use Pattern*	*Area (hac)*
	Total geographical area	2,54,381
	Forest	1,43,351
	Net cultivable area	76,025
	Area under different crops	
1.	Cereals and pulses	1,746
2.	Spices	4,132
3.	Fruits	1,609
4.	Vegetables	4,142
5.	Potato	5,357
6.	Oilseeds	120
7.	Flowers	37
8.	Otherfood crops	244
9.	Plantation crops	
	a) Tea	51,503
	b) Coffee	7,289
	c) Other non-food crops	1,837
	Total	76,025

Source: Status Report, Office of the JDH, Ootacamund (2001).

APPENDIX IV

Department of Agricultural Extension
and Rural Sociology,
Tamil Nadu Agricultural University,
Coimbatore – 641 003.

Dr.R.Ganesan
Professor and Head

Dear Sir/Madam

Mrs. C. Cinthia Fernandaz Ph.D scholar of this Dept. is undertaking a study entitled "***Participation of tribal women in agro-biodiversity conservation and postulating environmental education module for sustainable livelihood***" as part fulfillment of her Ph.D programme under my guidance. As part of her study she is developing an index to measure the livelihood sustainability of the tribal women as dependent variable.

In this connection fourteen components of livelihood sustainability of tribal women have been identified and presented in Annexure-A and the independent variables in Annexure-B based on the review of literature and discussion with experts. Considering your high academic qualifications and vast experience, we are pleased to seek your valuable judgement on the relevancy of these components. The definitions of each of these components are also provided to facilitate your judgements.

Kindly record your judgement in the three-point continuum 'relevant', 'some what relevant', 'not relevant' by putting (") mark in the appropriate column. If you feel any more important variables has left out kindly add the same with your judgement.

Thanking in advance for your kind contribution in completing this portion of her research work.

The schedule duly filled shall be mailed back at your earliest convenience in the enclosed self-addressed stamped envelope.

With kind regards.

Yours Sincerely

(R. Ganesan)

Encl:

1. Schedules –Annexures A and B
2. Self addressed, stamped envelopes

Participation of tribal women in agro-biodiversity conservation and postulating environmental education module for sustainable livelihood

Objectives

1. To study the perception of the tribal women on agro-biodiversity conservation.
2. To examine the participation of the tribal women in agro-biodiversity conservation practices
3. To analyze the livelihood sustainability of the tribal women through agro-biodiversity conservation.
4. To design an Environmental Education Module (EEM) for sustainable livelihood through agro-biodiversity conservation

Annexure A

Livelihood sustainability has been operationalised as the capacity of the tribal women to satisfy their basic needs, from the resource base available and how far it could fulfill their needs through the management and conservation of the natural resource and the Agro-biodiversity. This livelihood sustainability conserves land, water, plant and animal genetic resources. It is environmentally sound, technically appropriate, economically viable and socially acceptable.

Kindly record your judgement in the three-point continuum 'relevant', 'some what relevant', 'not relevant' by putting (") mark in the appropriate column.

Sl.No.	*Components*	*Relevancy*		
		R	*SWR*	*NR*
1.	**Food security:** Food is the basic need of the needs. How far the natural environment and the crops satisfy the food needs of the community.			
2.	**Shelter acquaintance:** The products of the forest or the biodiversity are used for building shelters. The feasibility of such produces is measured.			
3.	**Employment assurance:** The type of employment activities they are engaged for their daily livelihood.			
4.	**Income generation activities:** Income generation is an important economic activity through which their financial needs are satisfied.			
5.	**Environmental soundness**: Refers to how far the livelihood sustainable practices enriches the environment by conserving land, water, plant and animal genetic resources or at least does not harm the environment.			
6.	**Natural resource management:** Natural resources are the important criteria for livelihood, and how they are mamnaged and conserved for their own use is measured.			
7.	**Intergrated water management:** Water as a resource should be tapered fully water saving and storage measures becomes very much essential			
8.	**Intergrated nutrient management**: Application of right quantity of both organic and inorganic sources including soil amendment at proper time, method and combination aimed at increasing the use efficiency of added fertilizer input, causing minimum damage to the resource base.			

Contd...

Contd...

Sl.No.	*Components*	*Relevancy*		
		R	*SWR*	*NR*
9.	**Intergrated pest management**: Management system in the context of the associated environment and the population dynamics of the pest species (insects, diseases and weeds) utilizing all suitable techniques and methods in compatible manner to maintain pest population at levels below the economic injury.			
10.	**Soil conservation:** Soil is a very essential resource for crop cultivation and other activities. The conservation of soil is important for livelihood.			
11.	**Agro-biodiversity Conservation:** The diverse biological and agricultural components always play a very essential role in livelihood aquiantance.			
12.	**Marketability:** Extent to which the produces are marketed at a place, time and price convenient to the farmer.			
13.	**Decision making pattern**: Refers to the nature of decision making (individual, joint of collective or community as a whole) that the family has resorted to.			
14.	**Government policies**: Extent to which various acts, legislation and policies enacted by central and state governments from time to time to influence the livelihood of the people.			
	Others please specify:			

R: Relevant; SWR: Some what relevant; NR: Not relevant

Annexure B

Y_1: Participation of tribal women in agro-biodiversity conservation

Y_2: Livelihood sustainability through agro-biodiversity conservation

Kindly record your judgement in the three-point continuum 'relevant', 'some what relevant', 'not relevant' by putting (") mark in the appropriate column towards the relevancy of dependent variable Y_1 and Y_2.

Sl.No.	*Independent Variables*	*Relevancy*		
		R	*SWR*	*NR*
1.	Age			
2.	Educational status			
3.	Occupational status			
4.	Marital status			
5.	Family type			
6.	Family size			
7.	Farming experience			
8.	Farm size			
9.	Annual income			
10.	Social participation			
11.	Mass media exposure			
12.	Extension agency contact			
13.	Personal localite			
14.	Cosmo politeness			
15.	Innovativeness			
16.	Economic motivation			
17.	Value orientation			
18.	Risk orientation			
19.	Scientific orientation			
20.	Rational orientation			
21.	Fatalism			
22.	Socio cultural linkage			
23.	Conservatism liberalism			
24.	Progressiveness- traditionalism			
25.	Intra tribal communication			
26.	Self reliance			
27.	Gregariousness			
28.	Religious belief			
29.	Family norms			
30.	Attitude towards Agro-biodiversity Conservation			
31.	Attitude towards Environmental Education			
	Others please specify:			

R: Relevant; SWR: Some what relevant; NR: Not relevant

Name and Designation

APPENDIX V
Relevancy Coefficient of Major Components.

Sl.No.	Indicators	Relevancy Coefficient
1.	Increase in production	0.58
2.	Income generation activities	0.65
3.	Increase in consumption	0.42
4.	**Food security**	**0.76**
5.	Processing	0.45
6.	Exchange activities	0.42
7.	Personal satisfaction	0.55
8.	**Nutritional security**	**0.75**
9.	Information generation	0.55
10.	Social sustainability	0.52
11.	Problem identification	0.58
12.	Problem solving	0.56
13.	Nutritional security	0.54
14.	**Health status**	**0.78**
15.	Savings of money	0.55
16.	Saving of resources	0.45
17.	Knowledge development	0.41
18.	Attitude change	0.46
19.	**Soil and Water sustainability**	**0.74**
20.	Use of locally available resources	0.55
21.	Organic production	0.54
22.	Motivation	0.45
23.	Improvement and development	0.58
24.	**Shelter acquaintance**	**0.72**
25.	Status change	0.54
26.	Access to resources	0.56
27.	Increase in yield	0.52
28.	Decrease of disease	0.55
29.	Financial freedom	0.54
30.	**Educational sustainability**	**0.78**
31.	**Assets development**	**0.73**
32.	Socio-economic development	0.64
33.	Empowerment	0.48
34.	Leadership development	0.42

Contd...

Contd...

Sl.No.	Indicators	Relevancy Coefficient
35.	Educational development	0.54
36.	Income and expenditure	0.55
37.	Capacity development	0.52
38.	**Economic sustainability**	**0.76**
39.	Institutional building	0.54
40.	Resource development	0.55
41.	Enterprise promotion	0.56
42.	**Cultural sustainability**	**0.72**
43.	Gender equality	0.58
44.	Democratic freedom	0.42
45.	Women empowerment	0.46
46.	**Assets development**	**0.78**
47.	Village organizations	0.43
48.	Institutional development	0.47
49.	Watershed management	0.41
50.	Improved financial assistance	0.45
51.	Micro financing	0.46
52.	Increased private sector	0.58
53.	**Employment generation**	**0.73**
54.	Improving linkages	0.54
55.	**Infrastructure and marketing**	**0.78**
56.	Value addition of the produces	0.65
57.	Storage facilities	0.54
58.	Increased infrastructure development	0.48
59.	Participatory problem solving	0.54
60.	Increased voluntary organisations	0.55

APPENDIX VI
Minor Components Identified and Selected after Factor Analysis.

(n=200)

Sl.No.	Name of the Minor Component	Components			Communalities
		Factor 1	Factor 2	Factor 3	
I. Major component–Food sustainability					
1.	Food is obtained from the diversity	0.849*	0.251	0.208	0.784
2.	Timely availability of food from the diversity	0.747*	0.391	9.087	0.722
3.	Adequate supply of food	0.759*	0.510	0.585	0.802
4.	Collection of food is done in groups	0.556	0.334	0.157	0.655
5.	Insufficiency of food is reduced	0.280	0.838*	0.378	0.845
6.	Community organizes food collection	0.546	0.100	0.243	0.850
7.	Food is adequate to all the members of the family	0.253	0.727*	0.298	0.745
8.	Food is easily available	0.327	0.775*	0.387	0.723
9.	Only the nature is the ultimate source of food	0.575	0.511	0.525	0.717
10.	Forest produces also add to the food supply	0.762*	9.942E-02	0.572	0.741
II. Major Component–Nutritional Security					
1.	Agro-biodiversity supplies the basic nutrition to the members	0.917*	0.412	0.121	0.841
2.	It supplies the essential mineral and vitamin nutrition	0.905*	0.210	0.358	0.819
3.	Natural food is not that much nutritional	0.211	0.450	0.344	0.889
4.	Basic diet is satisfied by the diversity	0.943*	0.189	0.621	0.663
5.	Nutrition is uniformly supplied to the members	0.943*	0.389	0.211	0.889
6.	Crops alone are nutritional in nature	0.314	0.251	0.236	0.674
7.	Nutrition is easily available	0.915*	0.294	0.113	0.836
8.	Livestock supports the nutritional supply	0.893*	0.287	0.621	0.823
III Major Component–Health Management					
1.	Healthy food is supplied by the diversity	0.753*	0.364	0.217	0.700
2.	Agro-biodiversity has good medicinal value	0.835*	0.349	0.344	0.818
3.	Ancestors were healthy due to natural food	0.142	0.591	0.255	0.686
4.	Forest produces have high medicinal value	0.855*	0.233	0.151	0.785
5.	Cultivation of medicinal plants is essential for health management	0.756*	0.112	0.166	0.585
6.	Traditional crops have more medicinal value	0.804*	0.324	0.215	0.751

Contd...

Contd...

Sl.No.	Name of the Minor Component	Components			Communalities
		Factor 1	Factor 2	Factor 3	
IV. Major Component–Shelter Security					
1.	Materials from the biodiversity is used for building house	0.745*	0.160	0.240	0.641
2.	It could be easily collected	0.787*	0.210	0.215	0.720
3.	Nature is a good shelter	0.489	-1.217E-02	0.209	0.839
4.	Traditional method is used for building houses	0.856*	0.148	0.194	0.818
5.	Houses built from the available materials are safety	0.823*	0.336	0.169	0.837
6.	Wild animals security is assured	0.472	0.286	0.225	0.710
7.	All the members could get adequate supply of materials	0.723*	-0.517	0.624	0.673
V. Major Component–Soil and Water Sustainability					
1.	Agro-biodiversity supplies the quality maintenance of soil and water	0.766*	0.318	0.350	0.782
2.	Traditional practices are essential for soil and water management	0.764*	0.429	0.682	0.828
3.	Agro-biodiversity sustains the adequacy of water	0.766*	0.318	0.682	0.809
4.	Conservation practices sustains the soil health	0.796*	0.376	0.259	0.780
5.	Soil and water is the only source of sustainable livelihood	0.678	-3.579E-02	0.211	0.575
6.	Soil and water conservation is essential for livelihood	0.856*	0.328	8.00E-02	0.846
VI. Major Component–Economic Sustainability					
1.	Agro-biodiversity provides economic sustenance	0.834*	0.510	10194	0.456
2.	Agro-biodiversity is Economically viable	0.826*	0.419	0.115	0.948
3.	Agro-biodiversity is sufficient for economic returns	0.467	0.490	0.171	0.660
4.	It provides monetary benefits to the residents	0.260	0.211	0.860*	0.854
5.	Traditional varieties have high economic value	0.705*	0.308	0.262	0.899
6.	Value addition of the produce is not necessary	0.696	2.288	0.265	0.835
7.	Forest produces have high monetary value	0.710*	0.445	0.419	0.811
VII. Major Component–Employment Generation					
1.	Biodiversity provides adequate employment	0.758*	0.301	0.285	0.951
2.	Community relies only on the Agro-biodiversity for employment	0.669	0.517	0.312	0.714
3.	Timely employment is provided	0.932*	0.524	0.224	0.934

Contd...

Contd...

Sl.No.	Name of the Minor Component	Components			Communalities
		Factor 1	Factor 2	Factor 3	
4.	Employment is provided to all the members of the family	0.855*	0.366	0.153	0.866
5.	Community organizations are essentials for generation of employment	0.200	0.628	0.236	0.884
6.	Employment gives mental satisfaction	0.842*	0.601	0.335	0.774
VIII. Major Component–Educational Security					
1.	Education is an important component in livelihood	0.712*	0.312	0.256	0.869
2.	Sufficient education is provided	0.759*	0.129	0.365	0.756
3.	Quality education is given	0.322	0.870*	0.235	0.893
4.	Continuous and lifelong education is possible	0.275	0.890*	0.125	0.739
IX. Major Component–Assets Management					
1.	Human capital asset (human resource development)	0.840*	0.402	0.221	0.867
2.	Socio-capital asset (social development/status)	0.911*	0.283	0.325	0.910
3.	Economic capital asset (economic development)	0.914*	0.318	0.215	0.925
X. Major component–Cultural Sustainability					
1.	Cultural heritage is maintained	0.818*	0.398	0.144	0.848
2.	Culture is enriched by the diversity	0.815*	0.305	0.245	0.818
3.	Cultural heritage could be transferred to other societies	0.432	0.448	0.244	0.919
4.	Cultural tradition is maintained by all the members	0.824*	0.308	0.162	0.800
5.	Cultural practices are always important for livelihood	0.739*	0.165	0.420	0.750
XI. Major Component–Infrastructure and Marketing Sustainability					
1.	Infrastructure is important for livelihood	0.750*	6.060E-02	0.235	0.770
2.	Value addition centers are essential	0.372	0.232	0.145	0.526
3.	Traditional crop varieties fetch good market price	0.911*	0.289	0.365	0.910
4.	Forest produces have good market value	0.914*	0.318	0.256	0.936
5.	Value added products could fetch more prices	0.333	0.213	0.264	0.553
6.	Good marketing practice fetch increased returns	0.730*	0.272	0.323	0.697

Contd...

Contd...

Sl.No.	Name of the Minor Component	Components			Communalities
		Factor 1	Factor 2	Factor 3	
XII. Major Component–Political Sustainability					
1.	Government policies are essential to maintain livelihood through Agro diversity	0.917*	0.236	0.532	0.841
2.	Political system encourages traditional practices	0.905*	0.156	0.321	0.819
3.	Indigenous management practices are essential	0.943*	0.233	0.222	0.889
4.	Government policies encourage the collection of forest produces	0.755*	0.214	0.344	0.836

APPENDIX VII

Participation of Tribal Women in Agro-biodiversity Conservation and Postulating Environmental Education Module for Sustainable Livelihood.

Interview Schedule

Part - I

Respondent No. :

Name :

Village Name :

Habitation :

1. Age :

2. Educational status

i) Illiterate:

ii) Primary education:

iii) Middle school education:

iv) Secondary School education:

v) Higher Secondary education:

vi) Collegiate education:

3. Experience in agro-biodiversity conservation : ____________ Years

4. Social participation

Sl.No.	*Particulars*	*Before*		*Now*	
		Member	*Office Bearer*	*Member*	*Office Bearer*
1.	Milk co-operatives				
2.	Co-operative bank/societies				
3.	Grama Panchayat				
4.	Panchayat Union				
5.	Farmers Discussion Group				
6.	Mahila Mandals				
7.	Youth club				
8.	Distinctive features (like MLA/MLC/MP)				
9.	Others, if any specify				

Contd...

Contd...

5. Information seeking behaviour

Give appropriate response for the following sources from which you seek information on Agro-biodiversity Conservation

Sl.No.	*Sources*	*Frequency*		
		Always	*Occasionally*	*Rarely*
1.	Friends			
2.	Relatives			
3.	Neighbours			
4.	Extension Workers			
5.	UPASI KVK scientists			
6.	NGO officials			
7.	Field officers of private agency			
8.	Radio			
9.	T.V.			
10.	Video			
11.	Newspaper			
12.	Wall paintings			
13.	Posters/Charts			
14.	Folders/Leaflets/Pamphlets etc			
15.	Demonstrations			
16.	Meetings/Discussions			
17.	Campaigns			
18.	Exhibitions			
19.	Tours/Field trips			
20.	Trainings			
21.	CD			
22.	Websites			
23.	e-journals			

6. Innovativeness

When would you prefer to adopt an improved technology?

i) As soon as it is brought to your knowledge

ii) After you have seen other farmers have adopted successfully in their farms

iii) You prefer to wait and take your own time

Contd...

Contd...

7. Economic motivation

Please state your agreement, disagreement and undecidedness on the following statements.

Sl.No.	Statement	SA	A	UD	DA	SDA
1.	Money alone does not give entire satisfaction in tribal women's life					
2.	A tribal women should adopt an innovation to get more money					
3.	The community gives due importance to the rich tribal women only					
4.	A tribal woman should give importance to social recognition rather than monetary recognition					
5.	Standard of living is more important than profit for the success of farm					
6.	To meet the goal of life money plays an important role					

8. Progressiveness

Please give your answers to the following questions.

Sl.No.	Questions	Yes	No
1.	A progressive farmer has to know about the recently developed scientific practices. Do you know about such practices?		
2.	A progressive farmer has to accept the improved practices which gives more income. a) Do you try to follow such improved recommended practices?		
	b) Do you accept new varieties/technologies when they are recommended?		
3.	When new varieties are recommended a progressive farmer has to help other farmers to adopt them.		
	a) Do you tell other farmers about the benefit of new varieties?		
	b) Do you help other farmers to follow the improved practices?		
4.	A progressive farmer is one who adopt the improved practices immediately (or) follow ahead of others.		
	a) In your village are you the first person to adopt the improved practices? If yes, specify them.		
	b) If no, have you adopted any new practices recommended in the last two years ahead of other farmer?		
5.	A progressive farmer should have contact with extension agency.		
	a) Have you approached the extension workers to get advise?		
6.	A progressive farmer should cultivate superior varieties or high yielding varieties.		
	a) Did you cultivate such varieties in the last year?		
7	a) A progressive farmer should fully follow the Agro-biodiversity Conservation practices		

Contd...

Contd...

9. Cosmopoliteness–Localiteness

Kindly state your agreement and disagreement on the following statements.

Sl.No.	Statement	SA	A	UD	DA	SDA
1.	A tribal woman can learn many things from the happenings and experience of her village stay					
2.	A tribal woman who has been something worked in her village, need not take any additional information from source outside her village					
3.	One can satisfy all her requirements out of her local resource available to her					
4.	Many things a tribal woman ought to know what happening may be of great advantages to a tribal woman. She therefore ought to know					
5.	At present, when transport and other communication facilities are developing a tribal woman should know more about things happening outside her village					

10. Attitude towards agro-biodiversity conservation

Sl.No.	Statement	SA	A	UD	DA	SDA
1.	Indiscriminate use of natural Biodiversity leads to Environmental Degradation					
2.	Pressing demand on Biodiversity will result in ecological problems					
3.	Lack of awareness about Agro-biodiversity conservation leads to ecodegradation					
4.	Agro-biodiversity degradation can be effectively controlled by conservation measures					
5.	Government policies help for Agro-biodiversity Conservation					

11. Decision making behaviour towards agro-biodiversity conservation

Please give your responses to the following question.

Sl.No.	Statements	Self	Consulting Spouse/ Elder	Consulting all Family Members	Help Others in Decision Taking	No Participation in Decision Making
1.	To select crops to be grown					
2.	To select new varieties/crops					
3.	Cultivating traditional crops					
4.	Adopting indigenous practices					

Contd...

Contd...

Sl.No.	Statements	Self	Consulting Spouse/ Elder	Consulting all Family Members	Help Others in Decision Taking	No Participation in Decision Making
5.	To adopt new practices					
6.	To change cropping pattern					
7.	To buy/sell land					
8.	Adopting bio-inputs					
9.	Marketing of the produce					
10.	Soil and water conservation practices					
11.	Land improvement					
12.	Establishing Seed banks					

PART B

I. Perception on agro-biodiversity conservation

A) Perception on *in situ* conservation

1. Maintenance of seeds of traditional crops in the field
   Easy/somewhat easy/difficult
2. Traditional crop cultivation
   Easy/somewhat easy/difficult
3. Livelihood sustainability through *in situ* conservation
   Supports/somewhat supports/least supports
4. Improvement in standard of living through *in situ* conservation
   Best improved/somewhat improved/least improved
5. Government policy for *in situ* conservation
   Supports/somewhat supports/least supports

B) Perception on *ex situ* conservation

1. Development of seed banks
   Easy/somewhat easy/difficult
2. Traditional food habits supports *ex situ* conservation
   Supports/somewhat supports/least supports
3. Cultural values safeguards Agro-biodiversity
   Supports/somewhat supports/least supports
4. Livelihood sustainability through *ex situ* conservation
   Supports/somewhat supports/least supports
5. Government policy for *ex situ* conservation
   Best suited/somewhat suited/not suitable

Contd...

Contd...

	II. Participation of tribal women in agro-biodiversity conservation			
Sl.No.	*Statement*	*Hpn*	*Mpn*	*Lpn*
	***In situ* conservation of agro-biodiversity**			
	A. Genetic resources			
1.	Cultivating domesticated crop varieties			
2.	Cultivating traditional crop varieties			
3.	Cultivating non domesticated wild varieties			
4.	Cultivating medicinal plants			
5.	Cultivating location specific crops like minor millets			
6.	Domestication of wild varieties			
7.	Collection of forest produce			
8.	Preservation of forests produce			
9.	Conserving soil quality and health			
10.	Water management practices			
11.	Livestock maintenance			
12.	Exotic breeds husbandry			
	B. Agro-biodiversity systems			
1.	Water shed management			
2.	Nutrient management for the crops			
3.	Organic cultivation			
4.	Using farm yard manure			
5.	Adopting natural management practices			
6.	Indigenous management practices			
7.	Using natural pollinators			
8.	Utilization of locally available resources			
	***Ex situ* conservation of agro-biodiversity**			
	C. Biotic factors			
1.	Gene banks			
2.	Seed banks of traditional crops			
3.	Medicinal plant garden			
4.	Kitchen gardening			
5.	Storage of seeds with advanced systems			
6.	Preservation of extinct varieties			
7.	Preserving soil biota			
8.	Conserving the micro organism population			

Contd...

Contd...

Sl.No.	*Statement*	*Hpn*	*Mpn*	*Lpn*
	D. Social values			
1.	Human participation			
2.	Easily available resources are used			
3.	Indigenous storage practices			
4.	Indigenous post harvest methodologies			
5.	Indigenous value addition of the produces			
6.	Community storage centres			
7.	Community crop management practices			
8.	Marketing of the produce locally and outside market			
9.	Exchange of seeds within and outside the community			
10.	Fixing price for the produce			
11.	Change in food habit			
12.	Diversification of agriculture			
13.	Soil and water testing labs			

III. Sustainable livelihood through agro-biodiversity conservation

Sl.No.	*Statement*	*SA*	*A*	*UD*	*DA*	*SDA*
	A. Food sustainability					
1.	Food is obtained from the diversity					
2.	Timely availability of food from the diversity					
3.	Adequate supply of food					
4.	Insufficiency of food is reduced					
5.	Food is adequate to all the members of the family					
6.	Food is easily available					
7.	Forest produces also add to the food supply					
	B. Nutritional security					
1.	Agro-biodiversity supplies the basic nutrition to the members					
2.	It supplies the essential mineral and vitamin nutrition					
3.	Basic diet is satisfied by the diversity					
4.	Nutrition is uniformly supplied to the members					
5.	Nutrition is easily available					
6.	Livestock supports the nutritional supply					
	C. Health management					
1.	Healthy food is supplied by the diversity					
2.	Agro-biodiversity has good medicinal value					

Contd...

Contd...

Sl.No.	Statement	SA	A	UD	DA	SDA
3.	Forest produces have high medicinal value					
4.	Cultivation of medicinal plants is essential for health management					
5.	Traditional crops have more medicinal value					
	D. Shelter security					
1.	Materials from the biodiversity is used for building house					
2.	It could be easily collected					
3.	Traditional method is used for building houses					
4.	Houses built from the available materials are safety					
5.	All the members could get adequate supply of materials					
	E. Soil and water sustainability					
1.	Agro-biodiversity supplies the quality maintenance of soil and water					
2.	Traditional practices are essential for soil and water management					
3.	Agro-biodiversity sustains the adequacy of water					
4.	Conservation practices sustains the soil health					
5.	Soil and water conservation is essential for livelihood					
	F. Economic sustainability					
1.	Agro-biodiversity provides economic sustenance					
2.	Agro-biodiversity is Economically viable					
3.	It provides monetary benefits to the residents					
4.	Traditional varieties have high economic value					
5.	Forest produces have high monetary value					
	G. Employment generation					
1.	Biodiversity provides adequate employment					
2.	Timely employment is provided					
3.	Employment is provided to all the members of the family					
4.	Employment gives mental satisfaction					
	H. Educational security					
1.	Education is an important component in livelihood					
2.	Sufficient education is provided					
3.	Quality education is given					
4.	Continuous and lifelong education is possible					
	I. Assets management					
1.	Human capital asset (human resource development)					
2.	Socio-capital asset (social development/status)					
3.	Economic capital asset (economic development)					

Contd...

Contd...

Sl.No.	Statement	SA	A	UD	DA	SDA
	J. Cultural sustainability					
1.	Cultural heritage is maintained					
2.	Culture is enriched by the diversity					
3.	Cultural tradition is maintained by all the members					
4.	Cultural practices are always important for livelihood					
	K. Infrastructure and marketing sustainability					
1.	Infrastructure is important for livelihood					
2.	Traditional crop varieties fetch good market price					
3.	Forest produces have good market value					
4.	Good marketing practice fetch increased returns					
	L. Political sustainability					
1.	Government policies are essential to maintain livelihood through Agro diversity					
2.	Political system encourages traditional practices					
3.	Indigenous management practices are essential					
4.	Government policies encourage the collection of forest produces.					

Part C

I. Constraints faced by the tribal women in participating in agro-biodiversity conservation practices

Sl.No.	Statement	Yes	No	Ranking
a.	**Technological constraints**			
1.	Complexity in practice			
2.	Non suitability of the technologies in the recent years			
3.	Lack of easily operating tools to conserve Agro-biodiversity			
4.	High risk and uncertainty in adoption of traditional crops			
	others			
b.	**Extension constraints**			
1.	Lack of awareness of recent technologies			
2.	Lack of training			
3.	Inadequate technical knowledge in Agro-biodiversity Conservation practices			
4.	Weak extension service			
5.	Lack of upgradation of knowledge in the Agro-biodiversity conservation practices			
	others			

Contd...

Contd...

Sl.No.	*Statement*	*Yes*	*No*	*Ranking*
c.	**Physical constraints**			
1.	Unfavourable climatic factors			
2.	Change in the climatic conditions			
3.	Change in soil conditions			
4.	Non availability of inputs/traditional seed varieties			
5.	Scarcity of labour			
6.	Shift in working conditions			
7.	Due to migration			
8.	Non availability of storage facilities			
	others			
d.	**Economic constraints**			
1.	Lack of credit to adopt the Agro-biodiversity Conservation Practices			
2.	High cost of plant protection chemicals			
3.	Delayed cash payment by middleman			
4.	High cost of labour			
	others			
e.	**Personal constraints**			
1.	Requirement of more physical strength to adopt the Agro-biodiversity conservation practices			
2.	Excessive burden of work			
3.	Illiteracy			
4.	Family norms			
	others			
f.	**Marketing constraints**			
1.	No market for traditional crops and location specific crops and varieties			
1.	Lack of marketing experience			
2.	Lack of marketing facility			
3.	Fluctuation in market prices			
4.	Exploitation by middlemen			
5.	Competition is high in the local market			
	others			

Contd...

Contd...

II. Suggestions for sustainable livelihood through agro-biodiversity conservation

Sl.No.	Statements	Yes	No
1.	Providing timely Environmental education		
2.	Providing training on recent technologies of Agro-biodiversity conservation practices		
3.	Imparting adequate knowledge on technological management		
4.	Exchanging knowledge by conducting meeting among groups		
5.	Ensuring technical guidance through efficient extension services		
6.	Providing inputs/traditional and location specific crops at subsidized rates		
7.	Fixing reasonable prices for produces of traditional crops		
8.	Developing tools and implements which can reduce the drudgery of tribal women in adopting Agro-biodiversity conservation practices		
9.	Periodical visits by extension functionaries		
10.	Creating awareness on environmental protection		
11.	Recruiting sufficient number of women extension personnel		
12.	Utilizing locally available resources		
13.	Family labour engagement for Agro-biodiversity conservation practices		
14.	Developing skilled labour through frequent trainings.		
15.	Arranging credit facility for cultivating traditional crops		
16.	Arranging marketing facility for traditional and location specific crops		
17.	Establishing small processing industries for seeds and produces.		
18.	Weather and market forecast		
19.	Encouraging indigenous practices		
20.	Encouraging establishment of seed banks.		
21.	Providing infrastructure for medicinal gardening		
22.	Providing community storage centers		
23.	Policy for traditional crop cultivating system		
24.	Others		

Contd...

Contd...

III. Mention some of the agro-biodiversity issues for environmental education

Sl.No.	*Issues*	*Preference*	
		Yes	*No*
1.	Seed banks		
2.	Medicinal plant garden		
3.	Indigenous practices		
4.	Soil conservation		
5.	Water management		
6.	Traditional crop varieties		
7.	Pollution control in Agriculture		
8.	Eco-friendly Agricultural Technologies		
9.	Organic cultivation		
10.	Importance of bio-inputs		
11.	Medicinal plant cultivation		
12.	Seed storage mechanism		
13.	Seed processing unit		
14.	Value addition centre		

APPENDIX VIII
Case Analyses.

Case studies are depth investigation of an individual to study their personal and integrated components that measure the success in the given situation. It provides a prelude to go for some remedial problems after identification of their actual needs. A successful case could be a model for the entire community.

Case I: Traditional seed banks-A way for agro-biodiversity conservation.

Name of the respondent: Mrs. Kundhidevi

Name of the village: Pudukothagiri

Name of the block: Kothagiri

Importance: Interested in traditional seed storage

Kothars are specially known for their cultural heritage and behaviour. They cultivate all kinds of hilly vegetables and plantation crops such as Tea and Coffee. This was a peculiar case of Mrs. Kundhidevi of the Pudukothagiri habitation in the Kothagiri revenue village. This village comprises of nearly 100 farm families all belonging to the Kotha community. The village was well designed with the temple in the right centre. The rows of the houses arise from the temple. All the houses were under the same pattern and tiled in nature. They had their own village leader and festivals. Although the community was tradition bound, nowadays they are moving out and more of urbanized in nature.

Mrs. Kundhidevi w/o Mr. Raman who was a bank staff, although he was working he had to resign his job due to his commitment as the village temple leader. This was the peculiarity of this community that they have to resign the job and take full time duty as the leader. Mrs. Kundhidevi was the responsible person in the family looking after the field and the cattle. She had two daughters studying in 10th and 8th standard in the government school in Kothagiri town.

She had completed forty years and had nearly 25 years of experience in farming in turn she highly participated in Agro-biodiversity conservation. She had completed her secondary school education in her own village in Pandalur Taluk of The Nilgiris. She was experienced in growing the traditional vegetable varieties such as potato, radish, beans, carrot, cabbage and beetroot. In the recent years some new varieties of potato, beans and carrot had been cultivated by her. But she greatly acknowledged the size and nutritive nature of the traditional vegetables grown. Particularly the colour of the carrot and beetroot was so dull in the recent modern varieties. It was so great in the traditional varieties. Even now she was conserving some of the traditional seeds of the vegetables. This was grown regularly and harvested in the month of January for their special festival occasion. In the recent years some piece of land had been occupied by tea plantation.

Mrs. Kundhidevi was an experienced person in the village for her knowledge on growing the traditional minor millets such as Thenai, Varagu and Samai. Still these millets occupy a major part of diet among the Kothars. The traditional crops played an important role in biodiversity conservation. She was an initiator among the villagers for self help group formation. She was the office bearer for gram panchayat and the district self help group. She was also the member of the cooperative bank. At present she had organized nearly 10 successful self help groups in the village. She had managed to make her group members actively participate in the group activities. They had availed loan from the nationalized banks and initiated their self employment activities like animal husbandry etc. The respondent has reported about one of her group members who had started to sign her own name after joining the group.

Contd...

Contd...

She would always rely on her relatives, neighbours and the extension workers for seeking information on Agro-biodiversity conservation. Some of the officers of the private seed company also approached for the help. Radio was her regular partner for gathering information and occasionally she relied on the TV and the newspapers. She frequently attended trainings and the demonstrations conducted by the development departments. Although she had attended some meetings and exhibitions of the local NGO, they were not of much use when considering Agro-biodiversity conservation. She was a medium level innovator because she adopted the improved technology after seeing other farm women adopting it successfully.

She had a strong agreement towards economic motivation of tribal women and says that money alone is not the entire satisfaction. She also confirmed that to meet any goal in life money plays an important role. She had medium level of progressiveness. Her level of cosmopliteness was high and her attitude towards Agro-biodiversity conservation was also high. When speaking about her decision making behaviour, she always consulted her family members for selection of crops, varieties and changing of cropping pattern. For adopting indigenous practices and establishment of seed banks she relied upon herself. She was maintaining a good seed bank in her house which comprising of the traditional seeds of vegetable crops such as beans, carrot and beetroot. She had some varieties of samai and thenai. She explained that traditional varieties were always superior to the modern varieties in quality, colour and size. Her perception towards Agro-biodiversity conservation was high. She strongly agreed that the Agro-biodiversity is well demonstrated in the traditional systems as mixture of crops and farming systems. She said that land races, traditional cultivars and farmer's varieties were sources of diversity in cultivated crops.

Considering the participation of Mrs.Kundhidevi in Agro-biodiversity conservation she said that she was more interested towards the *in situ* conservation practices. She had high experience in cultivating traditional varieties. She was a good conservator towards the location specific crops like minor millets. She was not much interested towards the water management and livestock maintenance. She believed that the forest produces also play an important role in biodiversity. She adopted the natural management practices and utilized the locally available resources. She had established a seed bank on her own which is an initiative for the biotic factors. She believed that indigenous management system alone could pave way for the biodiversity conservation. Community participation was also highly needed. She also exchanged the seeds within and outside the community.

She denoted that the livelihood could be sustained with the help of Agro-biodiversity in the long run. She argued that food, shelter and economic sustainability where the important components of sustainable livelihood through Agro-biodiversity conservation. Nutrition and health were the other important components of livelihood sustainability. When the cultural and traditional factors were maintained it established the livelihood of the individual which sustains the food habit and living of the individual. If the traditional foods were taken it will lead to the cultivation of minor millets.

She mentioned some of the major constraints faced by the tribal women in practicing the Agro-biodiversity conservation measures. Non availability of the technologies was the major technical constraints faced by the respondent. Inadequate technical knowledge in Agro-biodiversity conservation practices and unfavourable climatic conditions were the other important constraints. Lack of credit to adopt the practices and requirement of more physical strength were the other important constraints as mentioned by the individual. Development of community eco-clubs and providing timely environmental education to the people was a good suggestive measure to overcome the constraints. She also suggested that development of skilled labour through frequent trainings as another important measure to overcome the constraints. When mentioning about the Agro-biodiversity conservation measures which need Environmental education, she mentioned about the promotion of eco-clubs and establishment of community seed banks. Finally she suggested that group activities would always be more welcoming than targeting the individual.

Contd...

Contd...

Case II : Natural farming–The future of the agro-biodiversity

Name of the respondent: Mrs. Vasatha

Name of the village: Kunjapannai

Name of the block: Kothagiri

Importance: Interested in organic farming

Irulars are the predominant tribal groups of Kunjapannai revenue village. They mainly depend on Agriculture for their occupation. Women were mainly concentrating on Agro-biodiversity conservation and collection of forest produces. This case is about an intelligent lady who concentrated much on organic and natural cultivation by name Mrs. Vasantha. She had a diversified cropping system in her farm with Coconut, Banana, Coffee, Tea and vegetable crops. Most of the crops were traditional varieties and stored seeds. She was the head of her family with a son and daughter. She was fifty years old. Although she was an illiterate, she organized and, managed her farm and home. She had an experience of nearly 27 years. She actively participated in Mahila mandals and self help groups. Although the women of Irula community were conservative in nature, she was an exception and breadwinner of her family.

Mrs. Vasantha was a low level innovator. She would prefer to wait and take her own time to adopt an improved technology. She would contact her friends, relatives and neighbours to get information regarding Agro-biodiversity conservation. Radio and TV were rare source of information for her. She had high level of economic motivation. She also suggested that a tribal woman should give importance to social recognition rather than monetary recognition.

Regarding progressiveness she said that a farm woman has to know about the recently developed scientific practices. She had medium level of progressiveness. She also helped other farm women in adopting recent development and scientific practices in Agro-biodiversity conservation. She had good contact with the extension agencies. She had low level of Cosmo politeness and localiteness. She felt that she could satisfy all her requirements out of her local resources available to her. She was well known about the happenings outside the village. Her attitude towards Agro-biodiversity conservation was medium. She strongly agreed that Agro-biodiversity degradation can be effectively controlled by conservation measures and government policies has to help to promote Agro-biodiversity conservation practices. Whenever she wanted to take a decision regarding the conservation practices she always contacted her family members. To adopt new crops to change the cropping pattern, to sell/buy the land and to market the produces she always contacted other members of the society.

She perceived that Agro-biodiversity was the sum total of plant diversity in the biosphere and the subset that was related to Agriculture is Agro-biodiversity. She argued that Agro-biodiversity was the basis of stability in Agricultural production livelihood of farming households and communities also sustainable development. She had a field that demonstrated Agro-biodiversity such as mixture of traditional system crops and farming systems. She preserved many wild seed relatives of cultivated crops and ethnic Agro ecosystems.

She had high level of participation in the *in situ* as well as *ex situ* Agro-biodiversity conservation practices. Considering the genetic resources, she cultivated the traditional and non-traditional varieties. She had some of the location specific crops such as dwarf coconut, wild coffee and trees such as Jack fruit. She highly concentrated on the preservation of forest produces and conservation of soil health and quality. She was not that much interested towards animal husbandry. Considering the Agro-biodiversity systems, water and nutrients were the two important systems that were concentrated more. She utilized the locally available resources for adoption of natural management

Contd...

Contd...

and indigenous management practices. She gave due importance to the *ex situ* conservation also. Her work towards the maintenance of seed bank and preservation of extinct varieties was recognizable. Socio-economic factors also played an important role in the conservation process. Indigenous storage and post harvest methods were highly used by them. She had high contact with other people outside the community. Hence she exchanged the seeds and collects variation varieties. Price fixation for the produce was an important indicator of socio-economic factor.

How far the Agro-biodiversity fulfilled the need for sustainable livelihood had been very well explained by her. Food and shelter were the two important components as mentioned by Mrs. Vasantha. Adequate food was obtained from the diversity. Infact the natural food was always nutritious because it always supplied the essential nutrients and minerals. She argued that food, shelter and economic sustainability where the important components of sustainable livelihood through Agro-biodiversity conservation. Nutrition and health were the other important components of livelihood sustainability. When the cultural and traditional factors are maintained it established the livelihood of the individual which sustains the food habit and living of the individual. If the traditional foods were taken it will lead to the cultivation of minor millets.

She mentioned some of the major constraints faced by the tribal women in practicing the Argo biodiversity conservation measures. Non availability of the technologies was the major technical constraints faced by the respondent. Inadequate technical knowledge in Agro-biodiversity conservation practices and unfavourable climatic conditions were the other important constraints. Lack of credit to adopt the practices and requirement of more physical strength were the other important constraints as mentioned by the individual. Development of community eco-clubs and providing timely environmental education to the people was a good suggestive measure to overcome the constraints. She also suggested that development of skilled labour through frequent trainings as another important measure to overcome the constraints. When mentioning about the Agro-biodiversity conservation measures which need Environmental education, she mentioned about the promotion of eco-clubs and establishment of community seed banks. Finally she suggested that group measure would always be more welcoming than targeting the individual. So the Environmental education module should speak of the community and participatory approach to solve the problems of Agro-biodiversity conservation.

www.ingramcontent.com/pod-product-compliance
Ingram Content Group UK Ltd.
Pitfield, Milton Keynes, MK11 3LW, UK
UKHW041643031026
15587UKWH00001B/473

9 789351 302698